Peter Mischke

Film Formation

in Modern Paint Systems

Translated by Ray Brown

Peter Mischke: Film Formation in Modern Paint Systems

ISBN 978-3-86630-861-9

Cover: BASF SE, Ludwigshafen, Germany

Mischke, Peter
Film Formation in Modern Paint Systems
Hannover: Vincentz Network, 2010
(European Coatings Tech Files)
ISBN 3-86630-861-2
ISBN 978-3-86630-861-9

Please ask for our book catalogue
Vincentz Network, Plathnerstr. 4c, 30175 Hannover, Germany
T (202) 684-6630, F (202) 380-9129
E-mail: books@american-coatings.com, www.american-coatings.com

Layout: Vincentz Network, Hannover, Germany

ISBN 3-86630-861-2
ISBN 978-3-86630-861-9

European Coatings Tech Files

Peter Mischke

Film Formation

in Modern Paint Systems

Translated by Ray Brown

Peter Mischke: Film Formation in Modern Paint Systems

ISBN 978-3-86630-861-9

Foreword

This book is all about how coating materials, especially paints, form films. A student asked me recently: "How is it possible to write a book on a topic like that?" Her question was right on the mark. There can be no doubting that common textbooks on surface coatings devote very few pages to the topic of film formation. And yet, close examination shows that film formation is a theme that surfaces and resurfaces in the expert literature and in technical discussions on coating technology, without ever being specifically addressed. And that leads on to the problems inherent in tackling the topic. These include how to isolate the subject matter from the related areas of application methods, chemistry/formulation, and metrology, and the rest of current coatings literature; the extent to which the latest specialist advances be taken into account. And last but not least: A university teacher in a broad sweep of disciplines ranging from chemical principles to colloid and polymer chemistry through to the science of binders, adhesives and paint chemistry, my purpose in writing this book is to present the basics of film formation in a broad canvas while striking the right balance of topics and not delving too far into other fields. I have drawn on my own expertise and university teaching experience as well as a wealth of information from monographs directly accessible to me and periodicals such as "Farbe und Lack".

The book commences with a brief explanation of the main coating concepts, before presenting methods of application as the first step towards a finished coating. This is followed by the physical aspects of drying. The middle section of the book deals at length with fundamental polymer and physicochemical aspects. It is intended to enable readers who lack an academic background in chemistry to understand the more specific content of the later chapters, without first having to study a comprehensive textbook on the subject. The second half of the book covers the fundamental film-forming principles and coating systems. A brief overview of test methods used to study film formation is presented in Annex 4.

So who should read this book and why? The intended readership includes:

- third-level students of coating technology wishing to deepen and round out their knowledge of pure paint technology on one hand and pure paint chemistry on the other;
- newcomers and career-changers seeking a readable textbook that is not overloaded with individual facts and/or specialist material
- skilled personnel in paintshops wondering why the various effects they observe as they go about their daily work actually occur and how they might intervene to solve particular problems that arise;
- finally, the book may be particularly instructive to laboratory staff and technicians seeking to gain a deeper understanding of film-forming mechanisms.

My thanks to everyone from the Department of Coatings Technology at Niederrhein University of Applied Sciences who kindly provided thoughtful suggestions, information and hands-on support. I would also like to thank Ilia Korolinskij, a graphics expert, who converted numerous sketches into a print-ready form. Finally, I am forever indebted to my family for sacrificing a great deal of quality time together during the year-and-a-half which I spent writing the book.

Willich/Germany, August 2009
Peter Mischke

Peter Mischke: Film Formation in Modern Paint Systems

ISBN 978-3-86630-861-9

Inhaltsverzeichnis

Foreword **4**

1 Introduction **9**
1.1 Basic concepts 9
1.2 General composition of coating materials 11
1.3 General formulation data 12
1.4 Basic physical properties 14
1.4.1 Density 14
1.4.2 Viscosity and flow behaviour 14
1.4.3 Surface tension 15
1.5 Classification of film formation 16

2 Application methods **18**
2.1 Spreading, flow coating 18
2.1.1 Brush and roller 18
2.1.2 Roller coating 18
2.1.3 Curtain coating 19
2.1.4 Flow coating 19
2.2 Dip-coating (dipping) 19
2.2.1 Conventional dip-coating 19
2.2.2 Electrodeposition 20
2.2.3 Autophoresis 22
2.3 Spray-painting 22
2.3.1 Air-spraying 23
2.3.1.1 High-pressure spraying 23
2.3.1.2 Low-pressure spraying 24
2.3.2 Airless spraying 24
2.3.3 Hot-spraying 24
2.3.4 Spraying with supercritical carbon dioxide 25
2.4 Electropainting 25
2.4.1 Electrostatically assisted conventional spraying 26
2.4.2 High-speed electrostatic atomisers 27
2.4.3 Purely electrostatic methods 27
2.4.4 Special effects of electropainting 28
2.5 Other application methods for liquid and pasty materials 28
2.6 General application conditions 29
2.7 Powder coating application 29
2.7.1 Powder coating 29
2.7.2 Electrostatic powder spraying 30
2.7.3 Powder-sintering methods 32

3 Substrate wetting, levelling, sagging, edge pulling **33**
3.1 Substrate wetting 33
3.2 Levelling 36
3.3 Sagging 39

3.4 Edge pulling ... 41

4 Physical principles of paint drying and curing ... 42
4.1 Convection drying ... 42
4.2 Heat transfer by means of infrared radiation ... 45
4.2.1 Physical principles ... 45
4.2.2 Industrial infrared lamps ... 47
4.2.3 Radiation input into the object for coating ... 47
4.3 UV irradiation ... 50
4.3.1 General information on UV radiation ... 50
4.3.2 UV emitters ... 51
4.3.3 UV radiant intensity and dosage ... 53
4.4 Further industrial drying and curing methods ... 54

5 General principles of film formation ... 55
5.1 Polymers ... 55
5.1.1 Basic definitions ... 55
5.1.2 Homopolymers and copolymers ... 56
5.1.3 Average molecular weight ... 57
5.1.4 Basic types of polymer ... 58
5.1.5 Molecular coils ... 59
5.1.6 Intermolecular forces and aggregates ... 59
5.1.7 Polymer networks ... 60
5.1.8 Glass transition ... 62
5.2 Solvents and polymer solutions ... 66
5.2.1 Solvents ... 66
5.2.1.1 Definition and classification ... 66
5.2.1.2 Volatility of solvents ... 67
5.2.2 Polymer solutions ... 69
5.2.2.1 General ... 69
5.2.2.2 Affinity between polymer and solvent; and solubility parameters ... 70
5.2.2.3 Vapour pressure of a solvent in a polymer solution ... 72
5.2.2.4 Viscosity of polymer solutions ... 73
5.3 Aqueous-disperse systems ... 74
5.3.1 Basic terms and classification ... 74
5.3.2 True and colloidal solutions ... 75
5.3.3 Primary dispersions ... 77
5.3.4 Secondary dispersions and emulsions ... 78
5.4 Diffusion ... 78
5.5 Basic principles of organic reactions ... 80
5.5.1 General ... 80
5.5.2 Timing in organic reactions ... 81
5.5.3 Chemical kinetics ... 82
5.5.4 Polar reactions ... 84
5.5.5 Free-radical reactions ... 85
5.5.6 Pericyclic reactions ... 86

6 Physical drying ... 87
6.1 Physical drying from solutions ... 87
6.1.1 Solvent transfer from film to ambient air, and heat balance ... 87
6.1.2 Sequence of events during drying ... 90

6.1.2.1 Evaporation, diffusion and solvent retention 90
6.1.2.2 Influence of layer thickness 91
6.1.3 Solvent selection 92
6.1.4 Evaporation processes in aqueous systems 93
6.1.5 Film formers for physically drying paints 95
6.2 Physical film formation from dispersions 96
6.2.1 Waterborne primary dispersions 96
6.2.1.1 Qualitative aspects of film formation 96
6.2.1.2 Physical models 98
6.2.1.3 Minimum film-forming temperature and coalescing agents 100
6.2.1.4 Other ways of lowering the MFFT 101
6.2.1.5 Pigmented dispersion coating materials 102
6.2.2 Waterborne secondary dispersions 102
6.2.3 Emulsions 103
6.2.4 Non-aqueous disperse systems 104

7 Oxidative crosslinking 105
7.1 General information and types of binders 105
7.2 Mechanisms of oxidative drying 108
7.2.1 Isolated double bonds 108
7.2.2 Conjugated double bonds 110
7.3 Siccatives 111
7.4 Preventing skinning 113
7.5 Influences on oxidative drying 113

8 Curing of liquid coatings by step-growth reactions 115
8.1 General 115
8.2 Polyaddition and polycondensation 115
8.2.1 Formal principles of molecular enlargement and crosslinking 117
8.2.2 Fundamental physicochemical principles of crosslinking 121
8.3 Important crosslinking reactions 124
8.4 Selected crosslinking reactions in detail 130
8.4.1 Formation of polyurethane systems 130
8.4.1.1 Conventional 2-pack curing 130
8.4.1.2 Waterborne 2-pack curing 132
8.4.1.3 Moisture curing 133
8.4.1.4 Stoving of blocked polyisocyanates 133
8.4.2 Crosslinking of epoxy resins 135
8.4.2.1 Epoxy resins and their reactivity 135
8.4.2.2 Additional information about curing with amines 137
8.4.2.3 Waterborne epoxy-amine systems 138
8.4.2.4 Additional information on curing with polyanhydrides 139
8.4.2.5 Curing via the OH groups 140
8.4.3 Curing of resin polyols with formaldehyde condensation resins 140
8.4.4 Crosslinking of silicic acid esters and sol-gel materials 144
8.4.5 General information on high solids paints 145

9 Film formation by coating powders 147
9.1 General information on coating powders and how they form films 147
9.2 Binders and how they crosslink 148
9.2.1 Hybrid powders 148

9.2.2 Epoxy powders 149
9.2.3 Polyester powders 149
9.2.4 Acrylic powders 151
9.2.5 General information on gelation and crosslinking 151
9.3 Physico-chemical aspects of film formation 152
9.3.1 Physical principles 152
9.3.2 Effect of additives, matting 155

10 Film formation by polymerisation 157
10.1 General polymerisation mechanisms 157
10.1.1 Free-radical polymerisation 157
10.1.2 Cationic polymerisation 159
10.2 Radiation curing 160
10.2.1 Free-radical curing with UV radiation 160
10.2.1.1 Initiation 160
10.2.1.2 Free-radical polymerisation 163
10.2.1.3 Oxygen inhibition 165
10.2.2 Electron beam curing (EBC) 166
10.2.3 Ionic UV curing 167
10.2.3.1 Initiation 167
10.2.3.2 Polymerisation 169
10.2.4 Waterborne binders 170
10.2.5 Dual cure 171
10.3 Curing of unsaturated polyesters 171
10.3.1 Polymerisation 171
10.3.2 Initiation, acceleration 173
10.3.3 Oxygen inhibition 174

Annex 1: The Orchard equation 175
Annex 2: The WLF equation 178
Annex 3: The gel-point theory of Flory and Stockmayer 180
Annex 4: Test methods used in film formation 183

Author 187

Index 188

1 Introduction

1.1 Basic concepts

The act of applying a solid, adhering, organic chemical layer or film to any kind of surface or substrate is called **coating** (note that inorganic coatings, such as distempers, are not the subject of this book) and the outcome is properly called a coating system. Depending on the **coating material** employed, the coating may be decorative, protective or otherwise functional and may be a varnish, an emulsion paint, a floor-coating compound or a filler. Before and immediately after application, the coating material is a liquid, a paste, or a powder (as in coating powders). The transition to finished coating requires the film of coating material to solidify and is called **film formation,** or, from an application point of view, **drying**. These terms and others employed in coating technology are defined in the EN ISO 4618 and also in the German standard DIN 55945 (2007-03).

In order for a coating material to form a film at all, it must contain a substance which is capable of forming a film by itself, i.e. without the presence of other chemical components. Such substances are called **film-forming agents** or **binders.**

Older standards, such as DIN 55945 (1988-12 and earlier), clearly distinguished between the terms "binder" and "film-forming agent". However, this distinction never really took hold in practice, and seems to have largely been abandoned.

According to DIN 55945 (1999-07),
- the film-forming agent is "the binder which is needed in order that the film may form"

In EN ISO 4618, we find:
- binder is the nonvolatile part of a medium, where medium is defined as all constituents of the liquid phase of a coating material.

In keeping with general parlance, in this book we will use the term "binder" even when, according to the traditional definition, a film-forming agent is meant.

Binders generally, and in this book, are the most important class of substance present in all coating materials. Every coating material contains a binder, which frequently is a mixture of several substances (resins, etc). We can say therefore that:
- the binder determines the fundamental properties of the coating material and the coating.

The job of the binder (or film-forming agent) is to form a more or less solid film and, where other ingredients are present, to embed them or to bind them to each other. Binders are generally organic substances that vary considerably in molecular size. Depending on their composition, degree of dissolution/dilution or manufacturing process, they are called either resins, e.g. alkyd resins, epoxy resins, and melamine resins, which can be present in 100 % concentrated, dissolved or dispersed form or **polymer dispersions.**

Binders composed of small to medium-sized molecules, equivalent to (mean) molecular weights ranging from several hundred to 10,000 g mol^{-1}, must be chemically transformed into large molecules or molecular networks during film formation. This is called hardening

Peter Mischke: Film Formation in Modern Paint Systems

ISBN 978-3-86630-861-9

or curing. Binders composed of large or long molecules (macromolecules) can form sufficiently solid films without chemical reaction; essentially, the solvent or dispersing agent in the formulation simply evaporates during film formation and the long thread-like molecules become matted, rather like felt. This type of film-forming process is called **physical drying.** The qualifier "physical" must be used in this case because "drying" can also include curing. Most films which are formed by curing additionally undergo physical drying or surface drying, especially in the early stages. Figure 1.1 shows a schematic diagram of the two basic film-forming mechanisms.

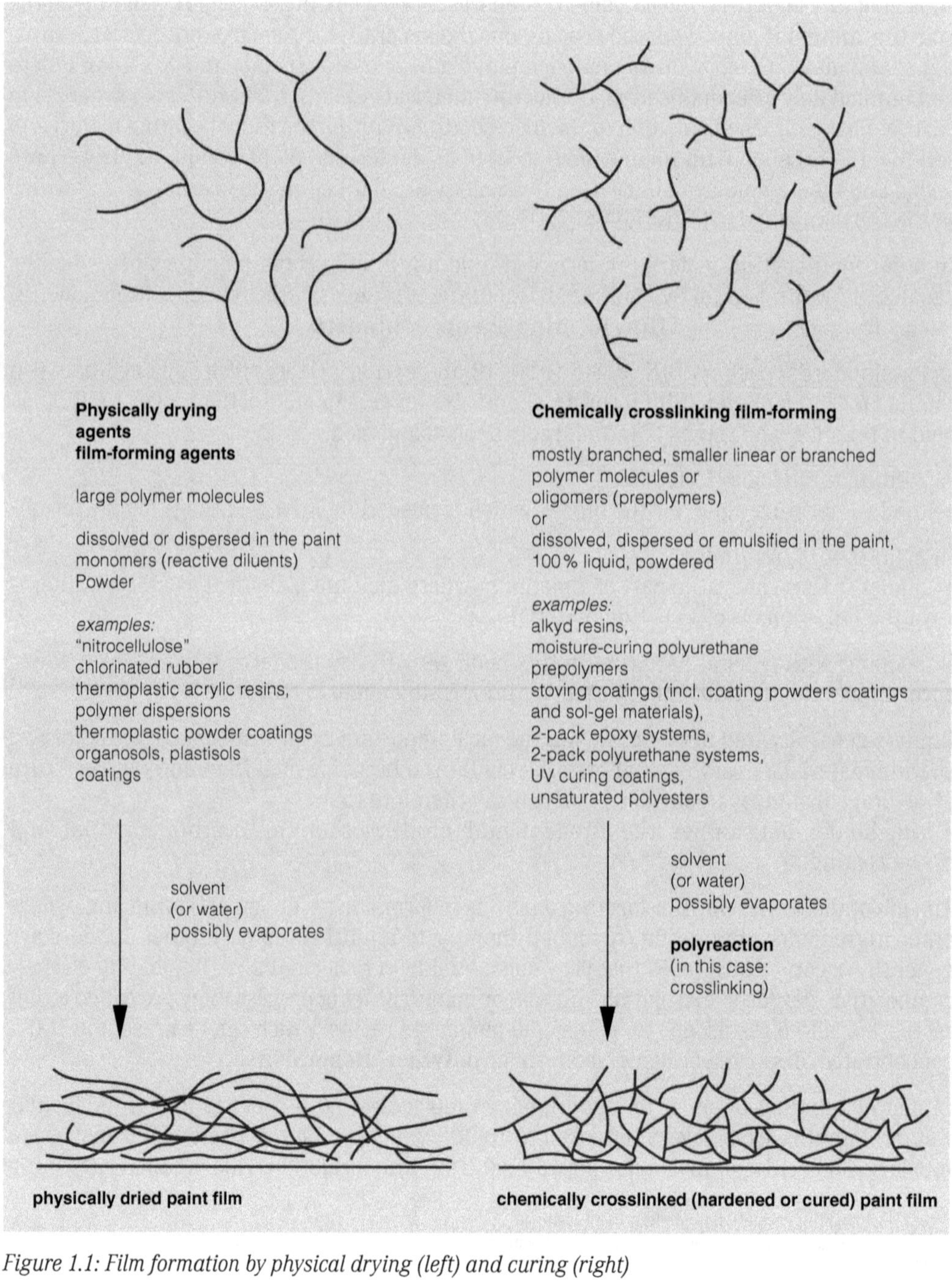

Figure 1.1: Film formation by physical drying (left) and curing (right)

1.2 General composition of coating materials

Once a specific coating material has been selected, it is a simple matter of referring to the instructions or technical leaflet to find out how it should be applied and dried. The leaflet or container will even contain general information about the chemical composition, possible health hazards and methods of disposal.

The composition of coating materials will be discussed in a very brief, generalised form below. Details and technical information are provided in this book as and when they are needed for an understanding of film formation. Table 1.1 presents an overview of the general composition of coating materials.

It should be pointed out that not all coating materials contain all classes of substance listed in Table 1.1. Thus, a clearcoat contains neither pigments nor extenders (in the classical sense), and a coating powder contains no solvents. Apart from binders, the ingredients mentioned in Table 1.1 will now be described in brief.

Pigments

These are very finely divided powders which are practically insoluble in the application medium, i.e. the liquid ingredients of the coating material, and/or act as colorants and/or prevent corrosion and/or have other functions. Examples: titanium dioxide (white pigment), carbon black (black pigment), quinacridone pigment (coloured pigment), pearlescent pigment (effect pigment), zinc phosphate (corrosion-protection pigment).

Fillers

Fillers, too, are powders which are virtually insoluble in the application medium. They impart volume or "build" to the coating material and provide or improve certain technological properties. They are typically used to modify flow properties, suppress cracking (provide reinforcement), improve ease of sanding, adhesive strength and weather resistance, matting, etc. The effect exerted by a filler depends critically on its particle size and particle shape (whether isometric, lamellar, or acicular). Examples: chalk, talcum, fibre fillers.

Additives

Additives, traditionally called "adjuvants", are substances which are added to the formulation in relatively small quantities and which, during manufacture and/or application of the coating material, engender or enhance certain properties or selectively improve the film properties. Additives vary in volatility, and will either remain in the coalescing film or escape from it more or less completely. Examples: Dispersing additives (nonvolatile), slip additives (nonvolatile), catalysts (mostly nonvolatile), defoamers (mostly nonvolatile), anti-skinning agents (volatile), film-forming agents (volatile).

Solvents

This term refers to liquids that are truly capable of dissolving a binder to create a molecular dispersion. Solvents (in the broader sense) are organic liquids, but can also be water. Examples: xylene (aromatic hydrocarbon), butyl acetate (ester), methyl isobutyl ketone (ketone), butyl glycol (ether alcohol), butanol (alcohol).

Table 1.1: General composition of coating materials

Coating material	
nonvolatile content	volatile content
binders	solvents and dispersing agents
pigments	volatile additives
film-forming agents	
nonvolatile additives	(usually volatile cleavage products during stoving)

If the solvent is added to the coating material to adjust its viscosity prior to processing, it is called a **diluent** or **thinner. Reactive diluents** are diluents which are almost completely chemically incorporated into the coalescing film and are therefore classed as binders.

Dispersing agents

Unlike solvents, these liquids (most often water) do not yield a molecular dispersion of a binder but rather contain it in the form of submicroscopic particles or droplets. The overall system is called a **dispersion** and – in contrast to a solution – usually has some cloudiness. A dispersing agent, too, may act as a diluent. Thus, it is quite common to lower the viscosity of an interior emulsion paint with a little bit of water for application by roller.

1.3 General formulation data

The simple paint formulation presented in Table 1.2 will now be used to explain the most important general data relating to formulations.

The **nonvolatile matter** (also called **solids content**) is the

- mass fraction of coating material remaining after a specified drying time.

The content of nonvolatile matter is determined in accordance with ISO 3251 by accurately weighing out, e.g., 2 g of the coating material into a lid, drying it under defined (or agreed) conditions, e.g., one hour at 130 °C in a paint drying oven, and then reweighing it. If chemical reactions during drying cause minor products to be released, the nonvolatile values can vary substantially with the temperature and duration of drying.

The theoretical content of nonvolatile matter in the formulation shown in Table 1.2 is calculated as follows:

$$60\,\% \cdot 0.75 + 27\,\% = 72.0\,\%$$

The additives are ignored if they are present in small quantities.

The content of nonvolatile matter is crucial for application because it indicates the mass of the film that will remain on an object after the coating has dried.

The **pigment-binder ratio** in the sample formulation is

$$27.0 : (60 \cdot 0.75) = 0.60 = 3 : 5$$

The pigment volume concentration, PVC, is the volume fraction of pigments and fillers in the dry film volume.

Table 1.2: Simple paint recipe (white house paint); from [3]

Ingredient[1)]	Mass fraction [%]
alkyd resin, 75 %, in white spirit	60.0
titanium dioxide (white pigment)	27.0
cobalat octoate (10 % Co)	0.2
zirconium complex (6 % Zr)	0.5
calcium octoate (5 % Ca)	1.7
white spirit	10.6
	100.0

[1)] *Figures quoted in percent are mass fractions (wt.%)*

The PVC is calculated as follows:

Equation 1.1:

$$\text{PVC} = \frac{V_p + V_F}{V_p + V_F + V_B}$$

V_P is the pigment volume, V_F is the filler volume, V_B is the binder volume

The binder volume here comprises the volume of all nonvolatile matter in the dry film without pigments and fillers, i.e. it also includes curing agents, reactive diluent and, perhaps, plasticiser.

The volumes have to be calculated from the mass fractions in the formulation and the corresponding densities, which are included in the leaflets supplied with the raw materials. In the example in Table 1.2, the density of the resin is 1.04 g cm^{-3}, and the pigment density is 4.1 g cm^{-3}. [1)]

The PVC is therefore:

$$\text{PVC} = \frac{\frac{27.0}{4.1} \cdot 100\,\%}{\frac{27.0}{4.1} + \frac{60 \cdot 0.75}{1.04}} = 13.2\,\%$$

The **critical pigment volume concentration, CPVC**, is the

- PVC at which the binder volume is just sufficient to fill the voids between the pigment and extender particles.

To calculate the CPVC, the size of the void volume of the pigment or pigment/filler in the formulation must be known. This is commonly measured by the **oil absorption value** (EN ISO 787-5), which is the quantity of linseed oil needed for converting 100 g of pigment or extender (or a mixture of both) into a coherent, non-lubricating paste. In practice, chemists determine these values direct from the binder solution or water used, instead of from linseed oil. Although oil absorption values have poor reproducibility, the oil absorption value remains important because of its direct use in calculating and developing the formulation.

The oil absorption value can be used to estimate the CPVC from the following, easily derived equation.

Equation 1.2:

$$\text{CPVC} = \frac{100\,\%}{1 + \frac{\rho_P \cdot \text{OA}}{100 \cdot \rho_L}}$$

ρ_P is the density of the pigment/extender
O_A is the oil absorption value of the pigment/filler
ρ_L is the density of the linseed oil (0.935 g cm^{-3})

From this formula, the CPVC of the example formulation is

$$\text{CPVC} = \frac{100\,\%}{1 + \frac{4.1 \cdot 20}{100 \cdot 0.935}} = 53.3\,\%$$

Where several pigments or extenders are present, the mean density of the mixture is used for ρ_P; the oil absorption value must be determined experimentally from the entire mixture, since the degree of packing (void filling) cannot be predicted from the individual data.

The **Q value** is the

- Quotient of PVC/CPVC (·100 %).

$$Q = \frac{13.2}{53.3} \triangleq 0.249 = 24.9\,\%$$

In the example, the Q value computes to

[1)] In this book, the power notation is used for composite units, e.g. J kg^{-1} K^{-1} (Joules per kilogram per Kelvin) instead of J/(kg K).

Formulations having Q-values of less than 100 % are said to be **subcritical**. They contain more binder than is necessary for filling the voids. The dry film is therefore not porous, i.e. it resembles a typical paint. Coating materials having Q values over 100 % are said to be **supercritical.** The films are porous, i.e. absorbent, and typical examples are interior emulsion paints and zinc dust paints. Figure 1.2 illustrates the qualitative relationship between Q value, coating properties and the type of coating material.

1.4 Basic physical properties

1.4.1 Density

The density, ρ, is the ratio of the mass, m, to the volume, V: $\rho = m/V$.
The SI unit is kg m^{-3}, but in practice the more instructive unit g cm^{-3} is used. Note that 1 g cm^{-3} = 1000 kg m^{-3}. An obsolete term for density is "specific weight".

1.4.2 Viscosity and flow behaviour

The science of flow is called rheology. The most important rheological parameter is the **dynamic viscosity**, η, of a flowable substance (fluid). It is defined as the ratio of **shear stress**, τ, to **shear rate** (velocity gradient) D or $\dot{\gamma}$ [2]:

Equation 1.3: $$\eta = \frac{\tau}{D} = \frac{\tau}{\dot{\gamma}}$$

The SI unit is Pa·s (pascal second) or one thousandth of it: mPa·s (millipascal second). It is related to the obsolete unit, cP, (centipoise), as follows: 1 cP = 1 mPa·s.

Rearranging the equation above to $\tau = \eta D$ makes the definition more instructive: the shear stress, i.e. the shearing force per unit area of the sheared layer, increases with increase in

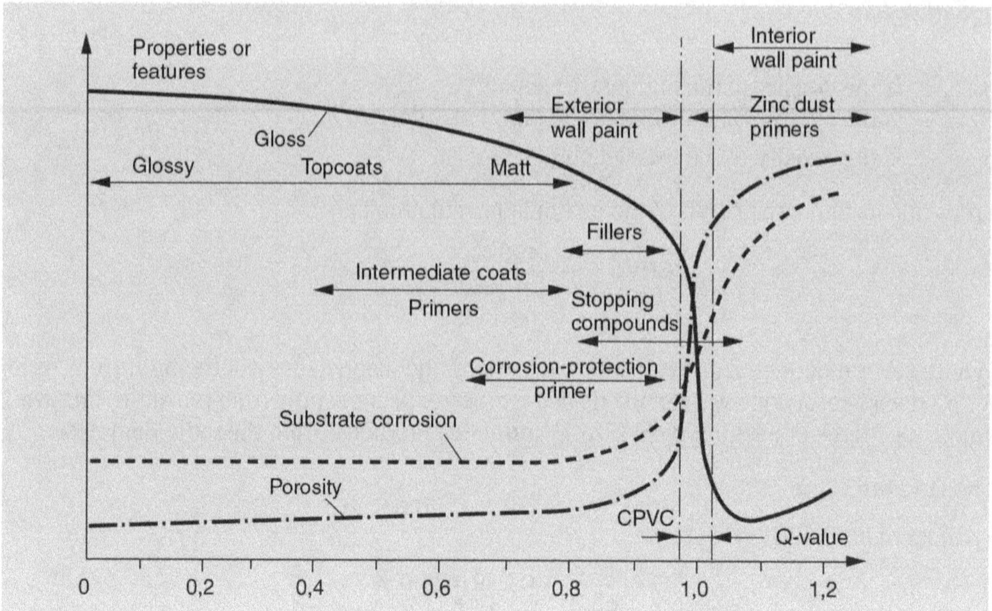

Figure 1.2: Various properties of coating materials, along with the corresponding Q value ranges from [3]

[2] A dot above a symbol indicates derivation by time (divided by time, temporal change).

viscosity of the material and/or in the rate of shearing. Figure 1.3 shows a thought experiment that can be used to define the dynamic viscosity (for further information, see "The Rheology Handbook" by T. Mezger).

In certain contexts, the **kinematic viscosity**, ν, is a useful parameter. It is the ratio of the dynamic viscosity to the density: $\nu = \eta/\rho$, and its SI units are $m^2\ s^{-1}$.

If the viscosity is independent of the shear rate, i.e. stays constant when the shear rate changes, the flow behaviour is said to be **Newtonian**. Coating materials usually exhibit more or less **non-Newtonian** behaviour, i.e. the viscosity is dependent on the shear and perhaps on the duration of shearing. The following types of viscosity are distinguished (see also Figure 1.4):

Structural viscosity (pseudoplasticity, shear thinning): the viscosity falls as the shear rate rises. The viscosity change is wholly reversible and responds almost instantaneously to the change in the shear rate.

Dilatancy (**shear thickening**) is the opposite of structural viscosity, i.e. the viscosity increases with increase in shear. Generally undesirable, dilatant behaviour is sometimes exhibited by very highly concentrated pigment or polymer dispersions.

Thixotropy exists when the viscosity at constant shear rate undergoes a fairly rapid asymptotic decrease and is restored when the shear stress is removed. Thixotropy originates from the reversible build up and degradation of loose **gel structures** in the fluid, and can be deliberately adjusted with appropriate **rheological additives.**

Rheopexy is the opposite of thixotropy, i.e. the viscosity increases under constant, weak shearing. Rheopectic materials are very rare.

A **yield point** is exhibited by a fluid which flows only after a minimum shear stress has been applied; it is nearly always observed in combination with structural viscosity or thixotropy. The presence of a yield point is also called **plasticity** and, if pronounced, leads to thickening. A yield point cannot be measured directly with simple viscometers, but rather only by extrapolation. Yield points are exhibited especially by products with high volume fractions of pigmentitious components, such as sealants and trowelling compounds, mastics, plastisols, emulsion paints and gels, whose non-sag properties can be seen with the eye.

The flow properties of coating materials are critical. For film formation, the two most important processes governed by rheology are levelling, which is generally desirable, and sagging, which is undesirable and takes the form of "runs" (or "tears", "curtains", etc; see Chapter 3).

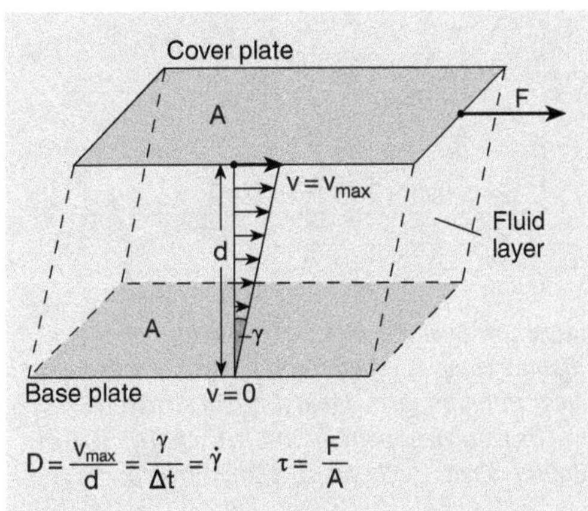

Figure 1.3: Thought experiment for defining dynamic viscosity

1.4.3 Surface tension

The surface tension of a liquid, σ_l, or a solid (material), σ_s, is the work per unit area dW/dA[3)] needed to enlarge the surface (at constant mass) by dA, where "surface" is understood to be the interface to the adjacent gas phase or vacuum. The SI unit is $N\ m^{-1}$ (or $mN\ m^{-1}$).

The word "tension" reflects the fact that the definition of σ above is based on another force-based equivalent:

$$\sigma = \frac{dW}{dA} = \frac{F\,ds}{L\,ds} = \frac{F}{L}$$

The length L is called the edge line. Figure 1.5 shows schematically how the surface of a liquid lamella stretched in a wireframe can be enlarged. Downward displacement of the wire bracket by ds incre ases the surface area by dA = 2 L ds (front and back). At the same time, the lamella pulls the wire bracket up to the top with a force F = 2 σ L. Thus, the external work to be done is dW = F ds = (2 σ L) ds = σ (2 L ds) = σ dA, and this is present in the surface energy of the added surface dA.

For practical purposes, we can conceive of the surface tension generally as a force that acts along every actual or imaginary line in or at the boundary of a surface. For the purposes of film formation, substrate wetting and flow depend on the surface tension of the coating.

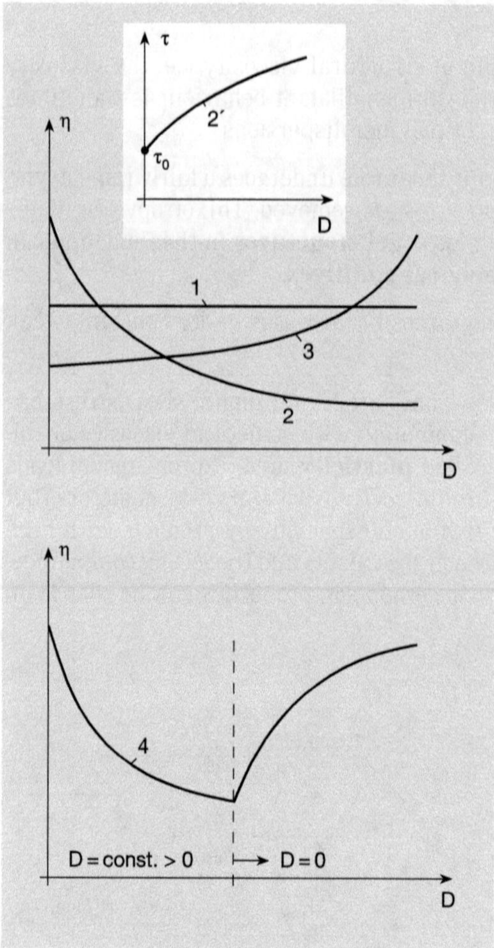

Figure 1.4: Basic rheological behaviour of coating materials: 1) Newtonian, 2) structurally viscous (shear thinning, pseudoplastic), 2') structurally viscous with yield point (in the τ/D chart), 3) dilatant (shear thickening), 4) thixotropic

Numerous other, often undesirable effects such as edge pulling, cratering and foaming, are also associated with surface tension (see Chapter 3).

1.5 Classification of film formation

Film formation – depending on the type of coating material – comprises various individual physical and chemical processes, which sometimes overlap and influence each other. The various types of film formation and drying undergone by coating materials can be classified as follows.

- **Physically drying coating materials**
  - solvent or dispersion-based (binders genuinely dissolved or dispersed)
  - film formation solely through evaporation of the solvent/water at different temperatures and, perhaps, air humidities
  - examples: cellulose nitrate, and chlorinated rubber paints, emulsion paints.
- **Chemically curing coating materials**
  - solvent-free or emulsion-free, yet flowable ("liquid, 100 % systems")

[3] Where appropriate or necessary, physical quantities and or changes therein are expressed directly in the form of the differentials (d .... or ∂ ...)

- film formation by spontaneous, chemical crosslinking reaction between **base paint** and **curing agent (two-component or 2-pack systems)** or through activation of crosslinking by means of heat, UV radiation or electron beams (1-pack systems)
- examples: 2-Pack materials based on urethane or epoxies for high-build coating in masonry and corrosion protection, heat-curing 1-pack polyurethane coating compounds, radiation-curing wood/furniture coatings based on acrylates.

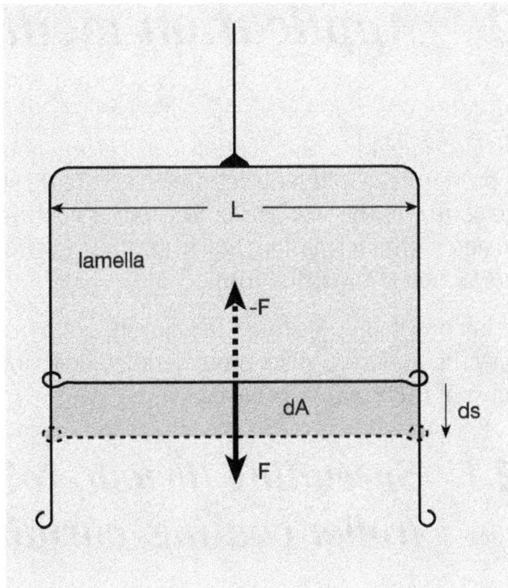

Figure 1.5: Surface tension of a liquid lamella in a wireframe (the area dA applies to the front and rear)

- **Physically (surface) drying and chemically curing coating materials**
  - binders genuinely dissolved or dispersed
  - film formation initially by partial evaporation of the solvent/water, but mainly by chemical crosslinking, cold-curing 2-pack or heat-curing 1-pack
  - examples: most industrial and automotive finishes.
- **Oxidatively curing or moisture-curing coating materials**
  - solventless or solventborne 1-pack coating materials that – perhaps after surface physical drying – crosslink with atmospheric oxygen or air humidity at room temperature or slightly above
  - examples: Alkyd resin house paints, moisture-curing polyurethane masonry paint.
- **Coating powders**
  - thermoplastic powders: film formation by purely physical fusion on the preheated objects
  - thermosetting (curing) powders: film formation by chemical crosslinking at elevated temperature after fusing and coalescence.

In summary, it should be noted that all types of drying entail an exchange of energy and, often, material with the surroundings. Thus, the external conditions during drying, i.e. the state of the environment around the object covered with the coating, determine the course of the film forming process. This particularly applies to the type and intensity of the energy input into the freshly applied layer (see Chapter 4).

2 Application methods

Strictly speaking, application of a coating material is not part of film formation, but it is both a prerequisite and a direct precursor. To be sure, the type of application and the resulting structure of the wet film and layer of coating powder have no effect on the basic mechanisms of film formation; however, they do affect the outcome of the coating process, i.e. the perfection of the dried film.

What now follows is intended merely as an overview of the application methods, with some specific remarks. Much more detailed descriptions can be found in [3, 7, 11, 15], upon which most of this chapter is based.

2.1 Spreading (brush, roller), flow coating (roller coating, curtain coating, dipping)

2.1.1 Brush and roller

Spreading of **paint materials** by brush still has its place in coating technology. The reasons are as follows:

- simple process
- high versatility as regards parts shape
- little or no masking of surfaces that are not to be coated
- good wetting of the substrate and incorporation of the coating into voids (cracks, holes, etc.)
- high **efficiency of application** (also known as **transfer efficiency**) [4)]
- wide range of materials processable without the need for precision adjustment

These are offset by weaknesses, such as

- low coverage rate
- uneven thickness
- uneven surface in the form of **brush marks**

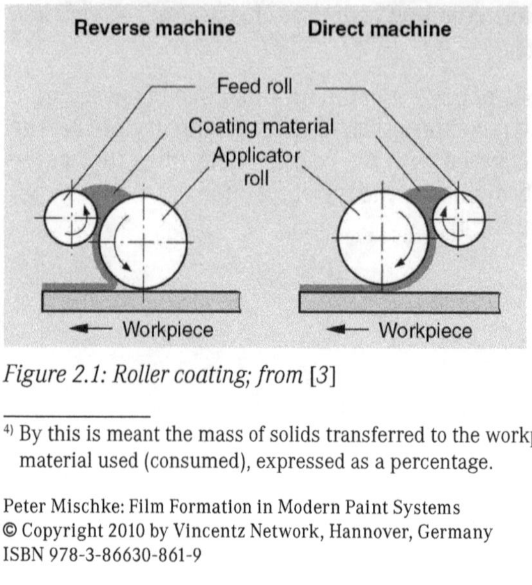

Figure 2.1: Roller coating; from [3]

Different types of **rollers**, such as long-haired or short-haired nylon wool or foam rollers yield much higher coverage rates and smoother surfaces. However, their use is largely limited to flat surfaces.

2.1.2 Roller coating

In roller coating the material is applied by mechanical rolls at a correspondingly high coverage rate. The coating material is applied to flat, panel-like

[4)] By this is meant the mass of solids transferred to the workpiece as a ratio of the total mass of solids of coating material used (consumed), expressed as a percentage.

Peter Mischke: Film Formation in Modern Paint Systems

ISBN 978-3-86630-861-9

or coil-shaped substrates by a more or less flexible roller. A distinction is made between classic **forward roller coating** and the more common **reverse** variant (see Figure 2.1). Whereas in forward roller coating, **film breakdown** between the roller and the substrate surface leads to **brush marks** that will not flow out completely where the coating thickness exceeds approx. 12 µm, the reverse variant can yield smooth coatings between 3 and 100 µm thick. All kinds of material viscosities can be present in reverse roller coating, too. Two rollers connected in series can be used to process two-component (2-pack) materials.

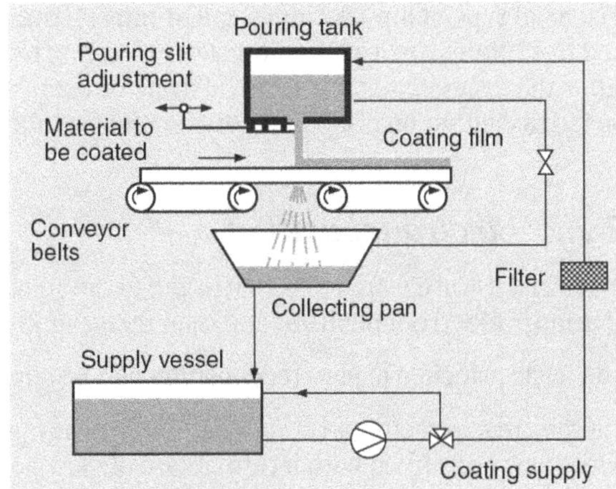

Figure 2.2: Curtain coating; from [3]

2.1.3 Curtain coating

In curtain coating, the paint falls through a horizontal slot as a very wide, thin curtain onto a virtually flat object carried on a divided conveyor belt. Excess paint is collected in a trough and is pumped back to the head (see Figure 2.2).

Curtain coating yields very flat films of 50 to 500 µm thickness. Since the paint only falls onto the panel with low kinetic energy, and is not, as it were, worked into the surface, good surface pretreatment or priming is essential for good wetting and adequate adhesion. 2-Pack coating is possible by arranging two heads in succession.

2.1.4 Flow coating

In flow-coating, the paint is either poured over or sprayed through jets onto complex and bulky articles, such as radiators, rough machinery parts, and roof tiles. The excess paint drains into a sump. To reduce inevitable sagging, the articles may be moved or rotated. This method is only suitable for low-quality, rough-and-ready coatings.

2.2 Dip-coating (dipping)

2.2.1 Conventional dip-coating

Conventional dip-coating does not make use of electrochemical processes to apply the paint to the surface of the object, unlike the case for electrodeposition (see Chapter 2.2.2) and autophoresis (Chapter 2.2.3). For dip-coating, coatings based on organic solvents can be used if provision is made for adequate encapsulation and extraction of fumes and explosion protection. Coatings based on non-flammable trichloroethylene are no longer in use for toxicological reasons. Advantageous for dip-coating is the use of waterborne paints, although there can be problems with foaming, bath instability due to pH shifts, and entrainment of contaminants (e.g. from pretreatment).

The comments made about flow-coating apply to wet-film quality and application areas. Unwanted sagging can be prevented by removing excess paint with high-voltage electrodes or by centrifuging.

A special type of dip-coating is called **barrelling**. In this, articles, such as buttons, hooks, and toy figures, are loaded into a slowly rotating barrel filled with a certain amount of paint. After the excess paint has drained away or been removed by centrifuging, the articles are heat-dried either on a separate wire screen or in the barrel itself.

2.2.2 Electrodeposition

Industrial use of electrodeposition began in the automotive sector in 1960s – firstly in the form of **anodic electrodeposition** and then increasingly as **cathodic electrodeposition**[5].

The basic principle of electrodeposition can be summarised as follows:

The pretreated metal parts for coating or priming are dipped in a sufficiently large tank which has been filled with a thin, relatively low-solids (10 to 20 % nonvolatiles) waterborne coating and are connected up as anodes (positive) or cathodes (negative) relative to the tank or separate counter electrodes. The voltage varies with the process and the time elapsed during the dipping phase from 150 to 400 V. A paint layer is deposited, which is almost dry and adheres strongly due to **electroendosmosis** (see below). The parts are removed from the tank, rinsed and stoved.

The primary electrochemical process in electrodeposition is electrolysis of water, which leads to the formation of protons and oxygen at the anode and hydroxide ions and hydrogen at the cathode.

- Anode: $2\,H_2O - 4\,e^- \longrightarrow 4\,H^+ + O_2$ $\quad (4\,H^+ + 4\,H_2O \longrightarrow 4\,H_3O^+)$
- Cathode: $4\,H_2O + 4\,e^- \longrightarrow 4\,OH^- + 2\,H_2$

The protons (or hydronium ions) formed at the anode migrate away from the surface into the bulk of the solution and cause the pH to fall by up to approximately 5 units at the **diffusion boundary layer**, which is a layer some 100 μm thick on the electrode. Accordingly, the pH at the cathode rises by about 5 units due to the formation of the hydroxide ions (Figure 2.3).

An anodic electrodeposition paint contains a resin bearing several carboxyl groups per molecule which are present in the bath as ions neutralised with amines and/or ammonia and so give rise to a dispersion of negative paint particles.

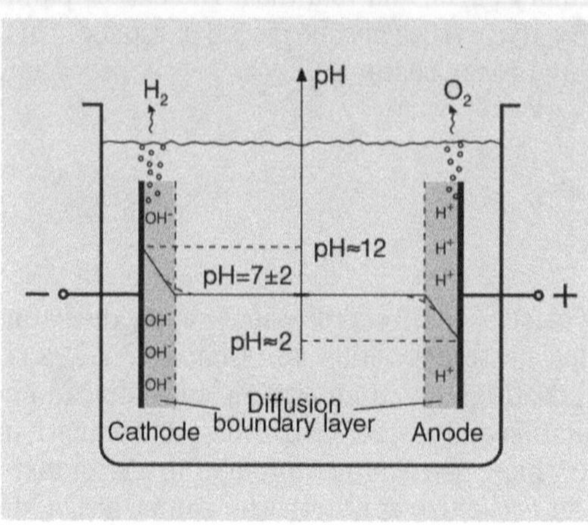

Figure 2.3: pH shift at the electrodes during the electrolysis of water (in anodic electrodeposition, the workpiece is the anode, while in cathodic electrodeposition it is the cathode)

Conversely, a cathodic electrodeposition resin is an amino oligomer which has been neutralised in the bath with lower carboxylic acids (mostly acetic acid), and forms positive paint particles (Figure 2.4).

As a result of convection in the bath, the negatively charged colloidal particles of the anodic electrodeposition coating enter the boundary layer of the positive anode to which they are attracted and at which they are protonated. Consequently, the resin molecules lose their ability to be diluted by water, and precipitation, also known as electrocoagulation, occurs on the substrate surface. Precisely the opposite

[5] Often incorrectly called electrophoretic coating or anaphoresis or cataphoresis.

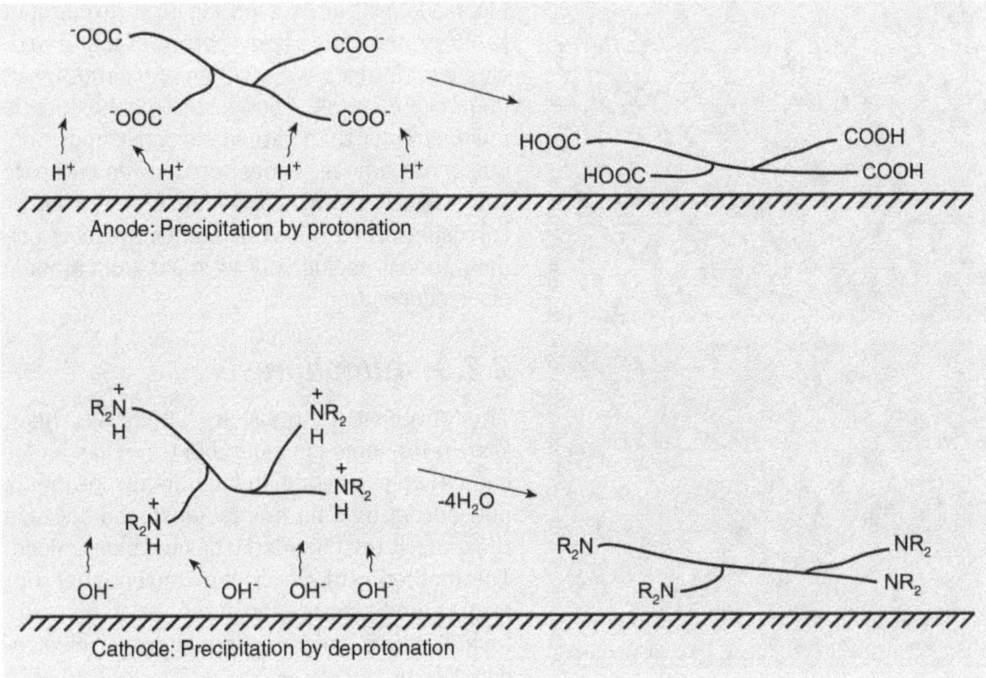

Figure 2.4: Neutralisation and precipitation (coagulation) of electrodeposition resins on the substrate surface

occurs in cathodic electrodeposition: the positive paint particles enter the cathode boundary layer, where they coagulate as a result of deprotonation (Figure 2.4). Since the deposited paint layer has poor electrical conductivity, the current quickly diminishes where most of the paint is deposited (at free surfaces facing the counter electrode). Layer growth comes to a standstill at approx. 20 µm thickness (value for cathodic electrodeposition), and more paint is increasingly deposited at electrically less favourable places, a phenomenon known as wrap around.

The main effects observed as regards film formation are:

Gas formation on the substrate surface causes blistering of the primary coating layers and renders them porous (see Figure 2.5). These structures must disappear during subsequent stoving. The layer is dewatered by electroosmosis (electroendosmosis). This is generally accepted (in simple terms) as the flow of electrolyte solutions through capillaries or porous bodies in an externally applied electric field. The effect stems from the fact that the protons (in anodic electrodeposition) and hydroxide ions (cathodic electrodeposition) which are emitted by the substrate electrodes entrain water envelopes with them as they pass through the paint. The solids content of the coating thus increases to up to approx. 90 %.

Connecting up the parts as anodes in anodic electrodeposition leads to significant formation of metal ions from the substrate, e.g. in the form of iron (III) ions, and maybe even to yellowing of the coating. Slight dissolution of aluminium or zinc in the form of hydroxo complexes has been demonstrated in cathodic electrodeposition.

For electrodeposition, as is the case for many conventional, higher-quality metal coating methods, the parts are usually phosphated, chromated, or otherwise furnished with organic **conversion layers**. These increase paint adhesion, and greatly improve protection against infiltration during subsequent corrosive attack. The conversion layers must be porous and/or electrically conducting enough to allow the high current flows needed for paint deposition.

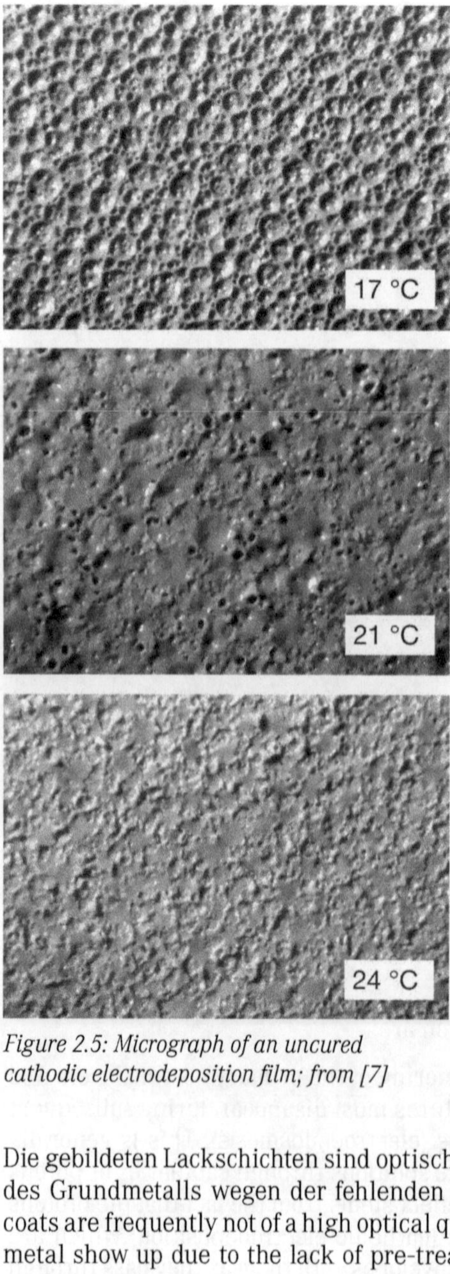

Figure 2.5: Micrograph of an uncured cathodic electrodeposition film; from [7]

Electrodeposition is used for high-throughput priming or single-layer coating of metal articles which are always given the same treatment (colour, etc.). Anodic electrodeposition is much simpler than cathodic electrodeposition, but is not now as widespread as its cathodic counterpart primarily because of the lower corrosion protection. Car bodies are now primed almost exclusively by means of cathodic electrodeposition.

2.2.3 Autophoresis

This dip-coating method, which has been known for some considerable time, has so far been unable to establish itself in Europe, but is now becoming a bit more widespread because the underlying chemistry has been optimised. The method ranks between conventional dip-coating and electrodeposition and its benefits include much greater simplicity over electrodeposition.

Low-viscosity liquid autophoresis paint is highly acidic and contains an acid-stabilised resin dispersion. Degreased, but otherwise unprocessed iron and steel parts are pickled by the acid after dipping. This causes the release of iron (II) ions, which in turn leads to destabilisation of the anionic dispersion and, in the presence of oxidants, oxidation to trivalent ions. The paint particles thus coagulate on the surface in the form of a porous and, owing to a lack of electrical osmosis, highly watery layer that is then thermally compacted and perhaps stoved.

Die gebildeten Lackschichten sind optisch oft nicht vollwertig, da sich alle Inhomogenitäten des Grundmetalls wegen der fehlenden Vorbehandlung abzeichnen. The resultant paint coats are frequently not of a high optical quality, because all the inhomogeneities in the base metal show up due to the lack of pre-treatment. Where this is not a problem, e.g., hidden surfaces, autophoresis is an attractive way of effecting one-coat painting of steel parts.

2.3 Spray-painting

Spray-painting methods can be divided into electrostatic and non-electrostatic variants, with the latter in some cases merely being an extension of the conventional methods (see Chapter 2.4). All spraying methods apply a cloud of droplets (spray mist) incompletely onto the object as a “mountain of droplets” (see Figure 3.3) which levels out quickly when the drops coalesce.

2-Pack coatings can be spray-applied as well. The base paint and hardener are either mixed prior to application or, where throughputs are high and/or the mixtures have short processing times, they are mixed *in situ* inside or in front of the gun with the aid of additional devices, such as static mixers ("Kenics" mixer) and metering equipment.

2.3.1 Air-spraying

2.3.1.1 High-pressure spraying

Conventional paint spraying with air at a pressure of between 2 and 8 bar (**high-pressure spraying**) is performed with a spray gun fitted at the front with an atomiser, as shown in Figure 2.6 (details may vary from brand to brand).

When the trigger is squeezed, compressed air from an air line is released from the centric, annular jet at the head of the gun, and rapidly expands. The negative pressure thereby created in front of the jet sucks and entrains the paint, which is emanating from the gun under slight pressure. The forces[6)] resulting from the exchange of momentum between the air flow and the paint flow cause the jet of paint to be atomised with a wide droplet size distribution. This spray jet now impinges at high speed (up to 30 m s^{-1}) on the object. For the most part, the larger droplets are deposited on the surface of the object. However, the finer ones are dragged away by the air current as it is deflected by, or sweeps past, the surface. This latter phenomenon is called overspray.

A quantitative description of the highly complex atomisation process is beyond the scope of this book. However, Figure 2.7 provides an overview of the influence of key parameters on the mean droplet diameter, the optimum value of which is 30 to 40 μm. The optimum initial viscosity of paint applied by high-pressure spraying is equivalent to a flow time of 20 s from a DIN-4 cup (DIN 53211, withdrawn in 1996) or 55 s from the ISO cup with 4-mm nozzle (DIN EN ISO 2431); this is equivalent to a kinematic viscosity of about $7 \cdot 10^{-5}$ m^2 s^{-1} or 70 mm^2 s^{-1}.

With regard to subsequent film formation, it must be remembered that the paint can lose significant quantities of volatile solvents through evaporation as it travels from the gun to the workpiece. The resulting rise in viscosity delays sagging, but unfortunately it also hampers levelling and penetration of porous surfaces.

High-pressure spraying yields the best spray results with simple technology. The disadvantage is its relatively low application efficiency, which is at most 50 % for large panels. It is still used for manually applying small quantities of paint that must yield a high-quality optical finish (e.g. automotive refinish paint).

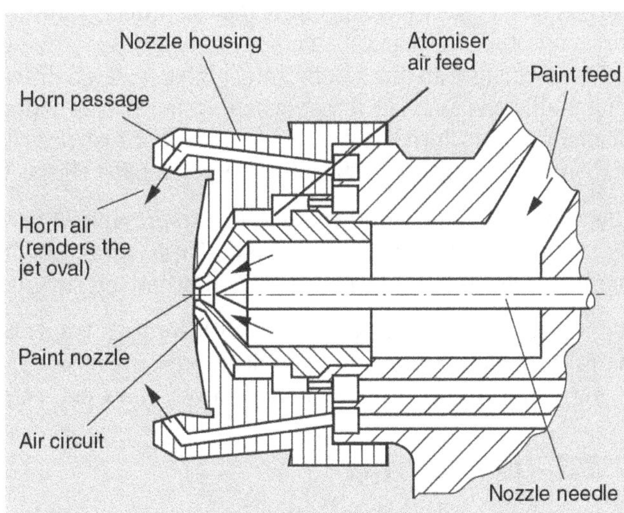

Figure 2.6: Cross-section through the atomising section of a high-pressure spray gun; from [3]

6) The force F is the temporal derivative of the impulse $(\vec{p} = m \cdot \vec{v})$: $\vec{F} = \frac{d\vec{p}}{dt}$

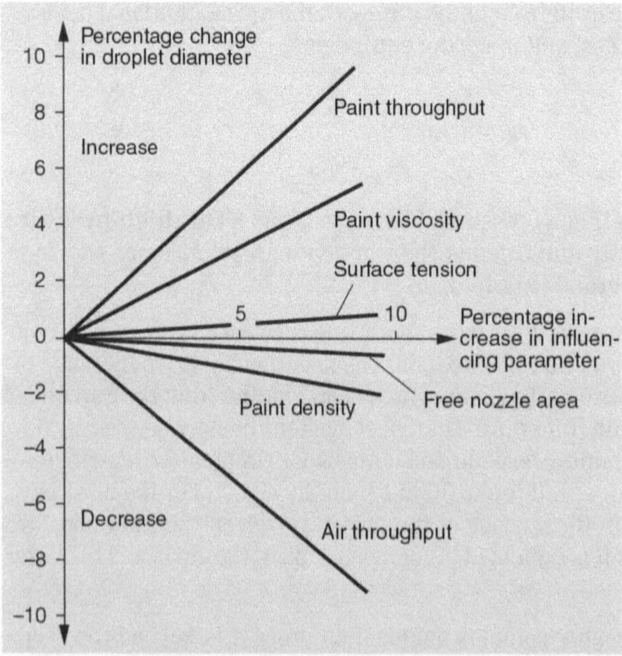

Figure 2.7: Influence of various parameters on the mean droplet diameter in high-pressure spraying; from [3]

2.3.1.2 Low-pressure spraying

If the compressed air used for air-spraying is at a maximum pressure of 0.7 bar, the volumetric flow rate at the gun head is correspondingly greater and so the spray mist has a lower content of fine droplets and there is less rebound from the workpiece. The result is up to 20 % greater application efficiency.

Also known as **HVLP** (high-volume low-pressure) spraying, this method, in conjunction with the latest spray-guns, yields a finish that almost matches that of high-pressure spraying, and it is becoming more and more popular.

2.3.2 Airless spraying

In this method, the material, which has a much lower solvent content and is thus far less viscous than in the case of air-spraying, is forced through a nozzle on the gun head at pressures of up to 400 bar. Due to high-energy turbulence in the paint and relaxation as it exits the nozzle, the paint jet is torn into fine droplets that impact on the workpiece as a fairly sharply defined spray jet.

Largely due to the lack of rebound of the voluminous spray mist from the workpiece that typifies air atomisation, the application efficiency of airless spraying is even higher than that of HVLP. Other advantages are the significantly higher material throughput and the possibility of processing undiluted paint. The disadvantage is the sharply defined spray jet and the greater droplet diameter, a fact which leads to a poorer quality finish (less flat and uneven). The airless method is therefore mainly used for coating (very) large flat objects where appearance is not so critical.

One way to combine the advantages of air-spraying with those of airless spraying is afforded by the **airmix method** (air-assisted airless spraying). It enables the spray jet to be optimally formed and the droplet size or, alternatively, the paint pressure to be reduced.

Airless spraying includes painting with aerosols (spray cans), in which a propellant forces paint through the spray nozzle under a pressure of several bars. It is mainly used in the do-it-yourself field. Recent developments even allow two-component coating out of a can.

2.3.3 Hot-spraying

Hot-spraying exploits the fact that the viscosity, η, of polymer solutions and thus of paints falls more or less in lines with **Arrhenius' equation** as the temperature rises:

Equation 2.1:

$$\eta = \eta_{\infty}\, e^{-\frac{E_{\eta}}{RT}}$$

η_∞ is the formal limiting value of η for $T \rightarrow \infty$
E_η is the molar activation energy of viscous flow
R is the general gas constant (8.314 J mol^{-1} K^{-1})
T is the absolute temperature (Kelvin temperature)

The change in paint viscosity with temperature rise follows the equation to a good approximation (see Figure 2.8).

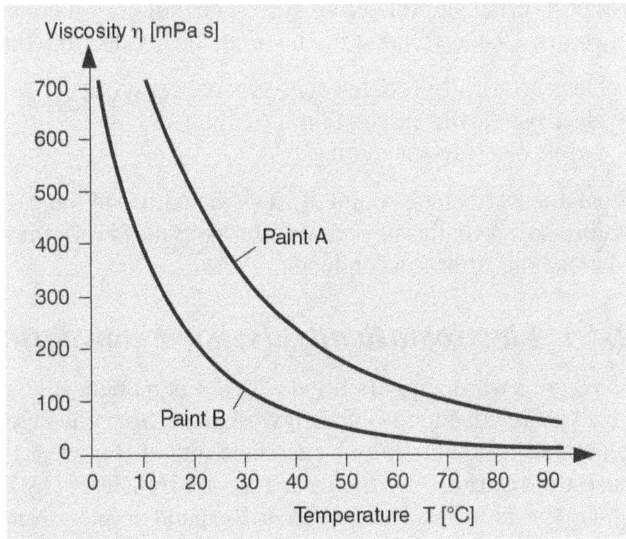

Figure 2.8: Viscosity of two paints as a function of temperature; from [7]

In practice, the hot-spray paint is heated in various ways to 60 to 80 °C before it enters the gun. Thanks to the viscosity drop, paints with higher solids content or lower solvent content can be sprayed direct, and so emissions are reduced. Cooling during travel to and on the workpiece causes the viscosity to increase rapidly again, and so sagging is largely suppressed, even in high-build coatings. Alternatively, a spray paint adjusted to normal viscosity can be applied at a lower air pressure, which leads to less overspray. There is also the possibility of additionally or exclusively using hot spray air, which leads to further advantages [24].

2.3.4 Spraying with supercritical carbon dioxide

Carbon dioxide gas (CO_2) reaches a **critical state** at 31 °C and 73 bar. At elevated temperature and elevated pressure, it enters the **supercritical state**, which means it behaves like a liquid and has good solvating power for binders. In a modified airless method, high-viscosity, very high-solids coatings can therefore be “diluted” with supercritical CO_2. Expansion of the CO_2 as it exits the nozzle leads to optimal droplet formation and, as in hot spraying, to a rapid increase in viscosity.

This so-called “**Unicarb**” method has very low emissions (CO_2 is not considered a **VOC** (volatile organic compound). Nonetheless, it has not become particularly widespread in Europe, probably because the results are not always satisfactory and the technical outlay is quite large.

2.4 Electropainting

The main weakness of conventional spray application, namely the poor application efficiency, can be mitigated by applying a high voltage field between the application member and the earthed workpiece such that the paint droplets, which are electrostatically charged during or immediately after spraying, are directed more strongly towards the workpiece than in the absence of electrical forces. In the vicinity of the object’s surface, there is an additional increase in field strength and thus in the attractive force between paint droplets and surface due to mirror charging induced in the object .

The basic principle of electropainting and the above-mentioned electrostatic effects are shown in Figure 2.9. Electropaint methods can be divided into the following main categories:

- electrostatically assisted conventional spraying
- electrostatically assisted high rotation and
- purely electrostatic methods

From an electrostatics point of view, the nature of the paint to be processed, whether it is a traditional solventborne or a waterborne type, is very important. The former is an insulator, whereas the latter is a conductor.

2.4.1 Electrostatically assisted conventional spraying

Spray guns whose tips are fitted with one or more needle-shaped high voltage electrodes (up to -90 kV relative to the earthed workpiece) are called **electrostatic guns**. Corona needles generate the conducting electric field and at the same time charge the paint droplets by **post-atomisation**, in which electrons are transferred by field emission from the needle tips, first to the air molecules and then to the paint droplets. Non-waterborne coatings can be charged additionally by **direct charging**, i.e. direct contact with the electrode, since the paint in the tube rules out the possibility of a short-circuit with the earthed paint delivery system.

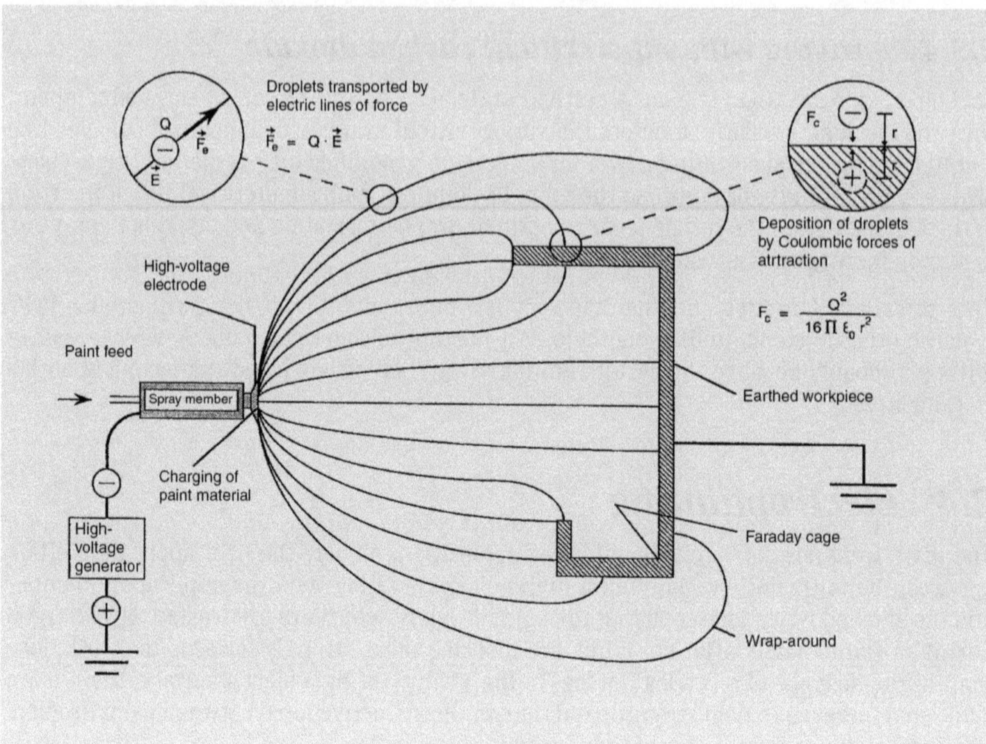

Figure 2.9: Basic principle of electropainting (simplified representation which ignores space charge effects caused by the charged air ions and paint droplets); from [15]

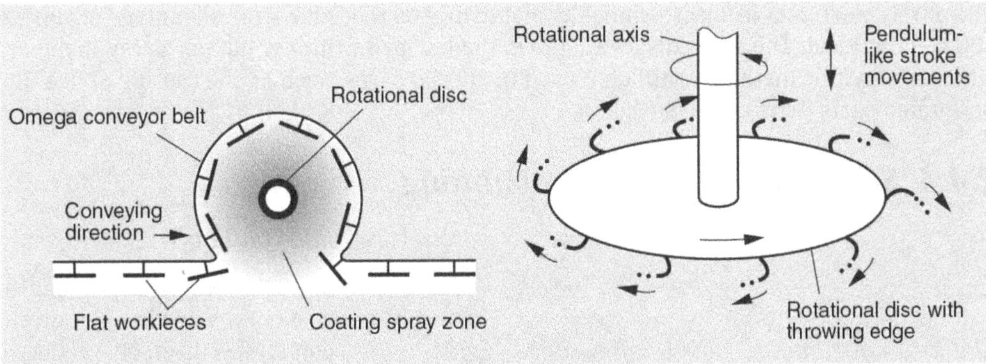

Figure 2.10: High-speed coating with a rotating disc in an omega loop; from [3]

2.4.2 High-speed electrostatic atomisers

Here, the paint is pumped continuously and centrically onto a **high-speed rotating disc** (diameter: 100 to 600 mm, speed up to 25,000 min^{-1}) or into a fairly small, high-speed rotating bell (diameter: 30 to 80 mm, speed up to 70,000 min^{-1}). The rotation of the atomiser throws the paint to the periphery where it is atomised under centrifugal force.

In the case of the disc atomiser, contact with the disc, which is under high voltage, charges the paint, which is then atomised and drawn by the electrostatic forces to the earthed workpieces, which are orbiting the disc in an **omega loop** (see Figure 2.10).

In the case of the bell atomiser, the spray mist is diverted towards the bell axis by an **air guide current** and additionally drawn towards the object by the electric field. Non-waterborne coatings are charged at the bell, but waterborne coatings (after atomisation) are usually corona-charged by a ring of external electrodes around the bell (see Figure 2.11). Alternatively, waterborne paints, too, can be direct-charged at the bell, provided that the bell and earth are **electrically isolated** via the paint column in the feed line to prevent a short-circuit.

High-speed rotating discs are generally encountered in industrial coating; one example is the painting of bicycle frames. High-speed rotating bells are mostly found in automotive OEM production-line painting.

2.4.3 Purely electrostatic methods

The sharp edges of a very fine slit nozzle (**electrostatic spraying slot**) or the edge of a relatively slowly rotating disc or bell can be used to atomise and spray paint. Due to contact with the edge, which is under high voltage, the high surface charge density on the paint leads to strong repulsive forces in the paint surface, which leads to characteristic **wavy sheet disintegration** and atomisation of the paint.

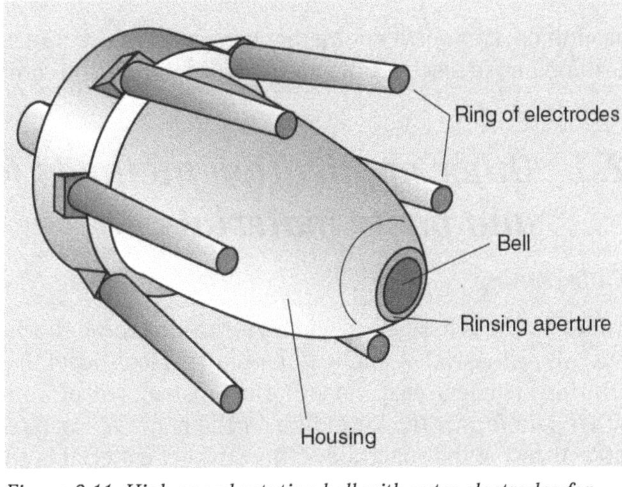

Figure 2.11: High-speed rotating bell with outer electrodes for atomising waterborne paint; from [3]

The purely electrostatic (electrostatically atomising) methods have the advantage of almost 100 % paint yield. The main disadvantage is the low paint throughput per spray member, which is why this method is still employed in special cases, such as the coating of heavily perforated parts (wire mesh, etc.).

2.4.4 Special effects of electropainting

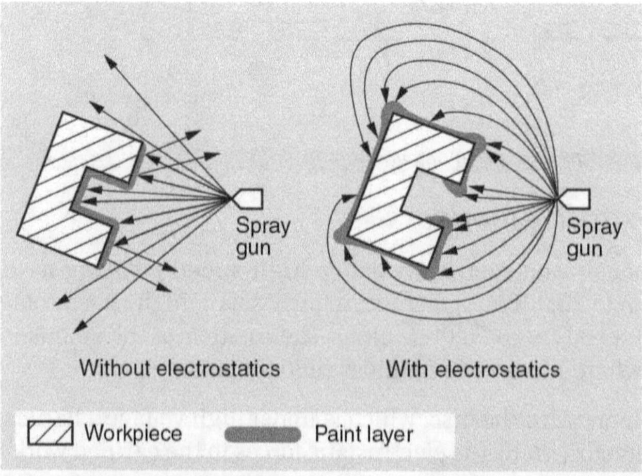

Figure 2.12: Deposition of paint droplets with/without electrostatic paint charging; from [3]

Edge build-up

Since the electric field strength on a curved surface at a given potential is inversely proportional to the curvature radius, it is particularly high at the edges and corners. Consequently, the paint droplets tend to concentrate there. If levelling occurs too slowly, the coating can build up at these points and eventually give rise to "picture framing" (see Figure 2.12).

Wrap-around

Usually a desired effect, which also occurs in electrodeposition, this is the tendency for paint to deposit on the rear of the workpiece (to a lesser extent), i.e. on the side facing away from the paint stream. Essentially, the droplets follow the magnetic field lines, some of which end up on the rear of the workpiece (see Figure 2.12).

Faraday cages

An electric field can only penetrate weakly into undercuts or even voids in the workpiece. Consequently, there is little or no electrostatic deposition of paint either (see Figure 2.12).

In addition, it is worth noting that these effects occur in an analogous manner in electrodeposition and in electrostatic powder spraying by the corona method (see Chapter 2.6.1).

2.5 Other application methods for liquid and pasty materials

Knife coating

Knife coating is basically a wiping technique through which excess material (coating, printing ink, adhesive) is applied to a flat surface (panel, flooring, paper, textile, plastic web, printing cylinders, etc.) and uniformly distributed, or worked in. A defined blade gap creates either a uniformly thick wet film, or the material is squeezed into the pores or depressions (structures) of the substrate as the surplus material is sheared off the surface. Very often, knife coating is also used in the laboratory to produce very flat films of defined thickness for test purposes.

In principle, knife coating is very similar to manual **filling**, i.e. application of trowelling compounds with a trowelling knife or similar tool.

Screen printing

Screen printing can be seen as an extension of knife coating. The relatively viscous (pasty) coating material or screen printing ink is forced by a knife through a textile fabric onto the substrate, and the fabric is removed. The alternating pattern of holes and non-holes in the fabric yields the corresponding structures on the substrate, with surface areas covered or not covered accordingly. In printed circuit board manufacturing, there is a purely technical purpose for this but in signage and poster printing it has informative and decorative purposes.

Vacuum coating

In vacuum coating, horizontal profiled strips or bars continuously pass through a vacuum chamber. The inlet and outlet apertures are only just a thin slit larger than the articles. Inside the chamber, as in flow-coating, the paint is sprayed onto the articles. When these emerge from the chamber, the excess paint is blown off by a stream of inflowing air.

Spin coating

The paint is metered centrically onto small flat articles (lenses, compact discs, etc.). The articles are then rapidly rotated about the vertical axis to yield a uniformly thick layer with good levelling properties.

2.6 General application conditions

As soon as a liquid coating material which contains solvent or dispersing agent (water) has been applied, it immediately starts to undergo physical drying. The rise in viscosity retards both undesirable sagging and desirable levelling. To just prevent the first and to maximise the second, the **evaporation rate (evaporation kinetics)** must be at an optimum. It is co-dependent on the temperature (of the object and the air), the air-flow and, in the case of waterborne paints, the **relative humidity**[7]: the dryer the air, the faster the water evaporates. Satisfying all these conditions creates an application window (see Figure 2.13).

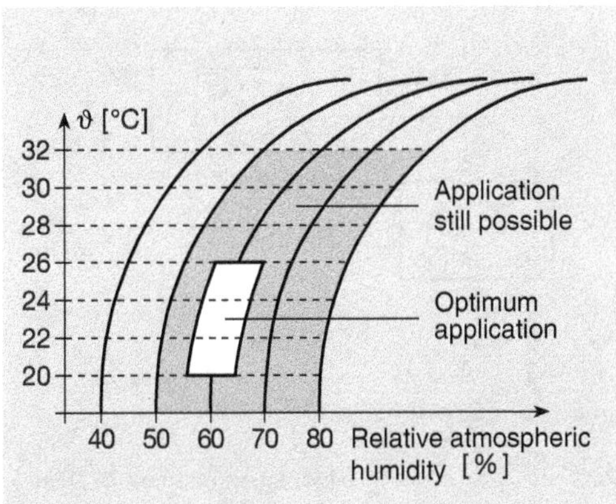

Figure 2.13: Application window for processing a waterborne paint; from [25]

2.7 Powder coating application

2.7.1 Powder coating

Powder coating are virtually emissions-free coating materials that can be processed with a transfer efficiency of some 95 %. They have proved to be an increasingly popular, very

[7] Relative atmospheric humidity: $\varphi = p / ps \cdot 100\ \%$; where p is the partial pressure of water vapour in air, ps is the saturation water vapour pressure

environmentally friendly alternative to conventional liquid coatings ever since their introduction in the mid-1960.

Powder coating generally consist of angular grains and are almost uniformly composed of binder(s) and additives. They mostly contain pigments as well and maybe some fillers. The mean particle size of stoving powders is usually between 20 and 45 µm, the particle size distribution ranging from 5 to 100 µm. Thermoplastic powders (fluidised-bed powders) are much coarser.

The powders are applied to the workpiece direct where, either immediately or in a subsequent curing process involving fusing, they level out to yield a smooth film, and may possibly undergo chemical crosslinking. Since the grains must fuse, only materials that at least offer short-term heat resistance, such as metals, glass, ceramics, thermosets and MDF (medium density fibre board), can be powder coated. Another limitation of powder coating is the minimum coating thickness of about 40 µm, because the grains on the surface must be close enough to prevent holidays in the coating.

2.7.2 Electrostatic powder spraying

In this method, the powder is fluidised by turbulent air and fed through tubes to a spray member, usually a gun or multi-gun configuration, although a powder-spray bell or rotary

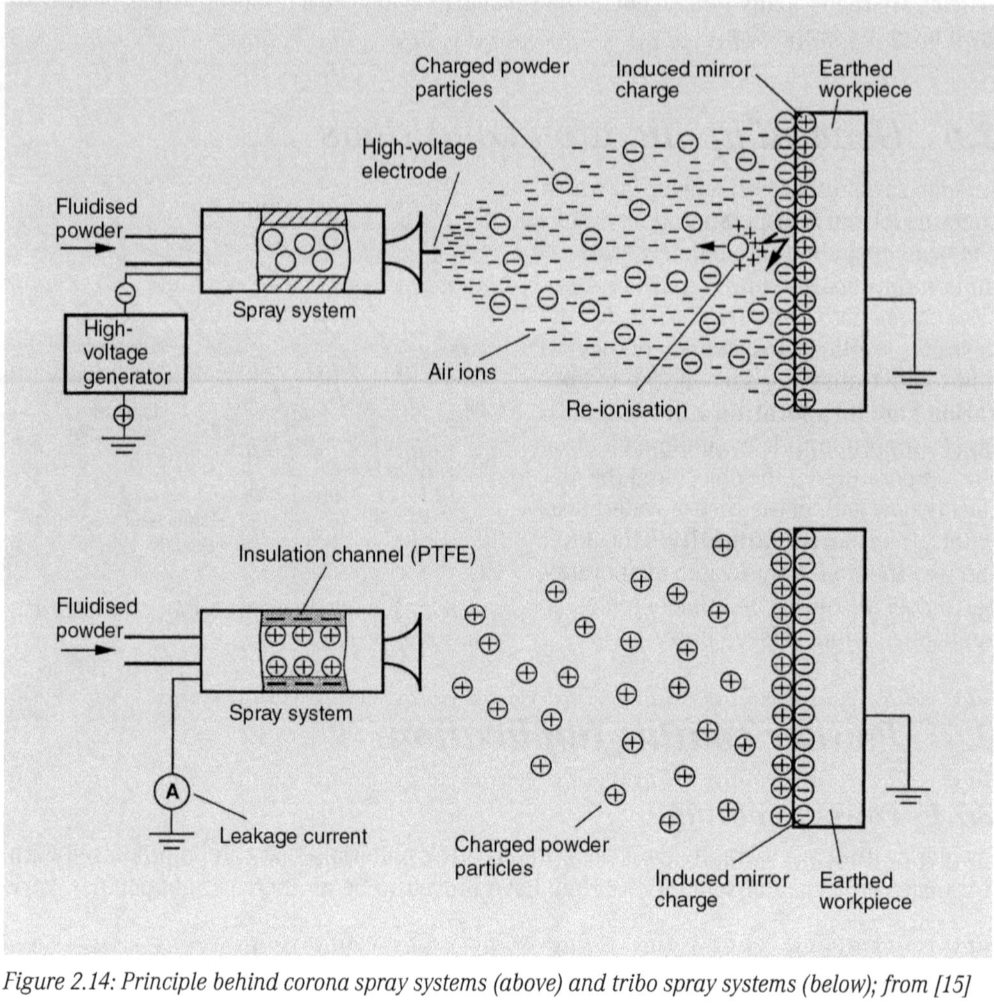

Figure 2.14: Principle behind corona spray systems (above) and tribo spray systems (below); from [15]

spray system is also possible. The powder particles become electrically charged and exit in the form of a cloud. Under electrostatic forces and carried along by the conveying air, auxiliary airflow and centrifugal forces, the particles impinge on the workpiece, where electrostatic forces of attraction (Coulombic forces) exerted by the induced mirror charges cause them to adhere to the substrate surface as a porous layer. Since the particles are insulators, they retain their charge in the layer as well. The overspray (comprising 50 to 70 %), which consists mostly of smaller particles, is collected, replenished with fresh powder and recycled to the spray member again, to yield a transfer efficiency of almost 100 %.

There are two ways in which the powder can be charged: **corona charging** and **tribo charging** (frictional charging). The two methods are shown schematically in Figure 2.14.

Corona charging

As in post-atomisation charging in liquid-coating spraying (see Chapter 2.4.1), when the particles exit the spray member, they undergo contactless charging by a corona electrode (90 kV max. for a hand gun and 120 kV for an automatic gun) and are transported by the externally applied field to the workpiece, where they adhere electrostatically as a layer in the manner described above.

Corona charging (provided that all the parameters are correctly set) is suitable for all Powder coating. Here, too, the existence of the strong external field gives rise to the electrostatic effects explained in Chapter 2.4.4. In addition, the strong flow of air ions through the layer into the earthed workpiece, and non-optimal technical parameters (see below), lead to back-spraying at high builds (from about 80 µm and above). Generally, the thickness of the powder layer is electrostatically limited to 120 to 150 µm.

Back-spraying is the phenomenon by which the powder layer becomes increasingly unstable as it grows thicker, with the result that powder particles suddenly start spraying back out of the layer, leaving craters or even holes behind them. One consequence can be that the finished powder coating has the texture of orange peel. Back-spraying is caused by an excessive accumulation of negative charge in the layer, mainly as the result of the stream of negatively charged air ions flying through the layer to the workpiece surface. Ionisation channels - i.e. electrical arcing - then occur in the layer. The particles either are discharged or become oppositely charge and fly out of the layer [13, 15].

Back-spraying can be minimised by lowering or varying the voltage as a function of workpiece shape, increasing the distance between the spray member and the workpiece, avoiding excessive layer thickness and possibly better earthing.

The various ways of achieving this technically nowadays are the use of low-ion spray systems with additional auxiliary electrodes, the use of spray members with variable current-voltage curve or the exclusive or additional use of tribo charging [15].

Tribo charging

The air/powder stream in the spray member (gun, finger nozzle etc.) is blown through a PTFE channel[8]. The particles exit the channel after having acquired a positive frictional charge through contact with the PTFE. Since no external electric field is present, the forces of attraction between the particles and the earthed workpiece are relatively weak coulombic forces. Near the workpiece surface, the attractive forces then become so strong on account of induction that the particles adhere as a powder coating.

[8] PTFE: Polytetrafluoroethylene, "Teflon"

Disadvantages or limitations of the tribo method mainly lie in the fact that not all powder coatings acquire a sufficiently high charge and that the transfer efficiency is lower than that of the corona method. However, the absence of the external field and the air-ion flow are advantages: layer thickness is more uniform over the entire workpiece, and recesses and even voids can be coated. No edge build-up occurs and back-spraying is much weaker or occurs only at very high layer thicknesses.

Electrostatic fluidised bed charging

This method is chiefly used for coating small, simple, earthed parts, which are passed through a more or less motionless cloud of charged powder that has been generated in a chamber or a basin by air jets or a fluid bed. Coatings ranging in thickness of 40 to 500 μm can be produced.

2.7.3 Powder-sintering methods

A shared feature of this group of methods is that a powder of a thermoplastic polymer, such as polyethylene and ethylene copolymer, polyamide, polyvinyl chloride, polyester or fluoropolymer, is applied to a workpiece which has been heated to 200 to 400 °C. The polymer generally melts to form a layer 200 to 600 μm thick. The heating temperature depends on the flow range (melting range) of the powder and the heat capacity of the workpiece.

The most important and oldest method is **fluidised-bed powder sintering.** It consists in immersing the hot part for 1 to 5 seconds in a basin containing powder which has been converted into a fluidised bed by air. In some cases, re-heating (post-gelling) is necessary for bringing the melting and levelling process to completion. Fluidised bed sintering is used for coating crockery baskets, fittings, garden furniture, etc.

Pipes, containers, small parts and large objects can be coated by other process variants which will not be discussed any further here.

3 Substrate wetting, levelling, sagging, edge pulling

3.1 Substrate wetting

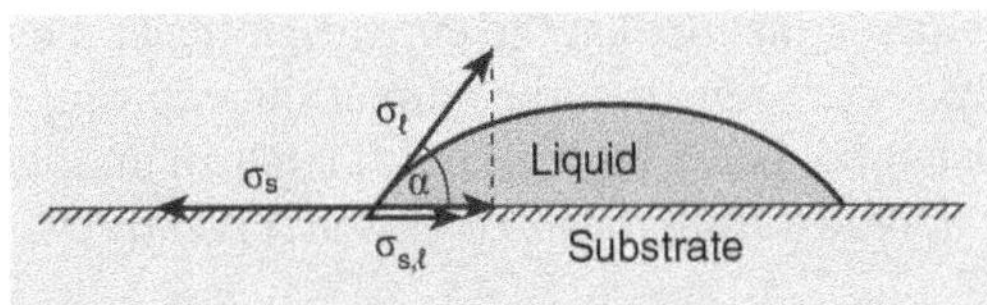

Figure 3.1: Balance of forces acting on a droplet on a solid surface. (Instead of the forces themselves, only the tensions are indicated)

The first condition governing the coatability (paintability, printability, etc.) of a surface (substrate) is that it can be wetted by the coating. To explain this concept, let us consider the simple model of a paint droplet on a substrate surface, as shown in Figure 3.1. The droplet is no longer a sphere, as it would be in a spray mist, but rather is deformed into a more or less flattened spherical segment by the force of gravity and by **surface** and **interfacial forces.** The circular boundary line between the droplet edge and the free solid surface forms an **edge line.** At any particular small section dL of this edge line, there are three forces acting, which – if nothing changes with respect to time – give rise to the following balance of forces (see Figure 3.1):

$$dF_s = dF_{s,l} + dF_l \cos\alpha$$
$$\sigma_s\, dL = \sigma_{s,l}\, dL + \sigma_l\, dL \cos\alpha$$

Dividing by dL:

Equation 3.1: $\sigma_s = \sigma_{s,l} + \sigma_l \cos\alpha$

σ_s is the surface energy (surface tension) of the substrate
$\sigma_{s,l}$ is the interfacial tension between liquid and substrate
σ_l is the surface tension of the liquid (paint)
α is the contact angle (wetting angle)

This is called **Young's equation** and it contains two unknowns, σ_s and $\sigma_{s,l}$. The **contact angle**, α, can be determined optically with appropriate **measuring instruments.** Measuring the surface tension, σ_l, of the coating material is also straightforward in principle [13].

We can therefore say that: the flatter the droplet,

- the smaller is the contact angle
- the greater is cos α (maximum 1)
- the better is the wetting of the substrate surface by the droplet.

Rearranging Young's equation, we get:

Equation 3.2: $$\alpha = \arccos\frac{\sigma_s - \sigma_{s,l}}{\sigma_s}$$

This shows that cos α for a specific interfacial tension $\sigma_{s,l}$ (rough approximation) increases (and hence the value of α itself decreases) with increase in σ_l and in σ_s. Finally, when the

Peter Mischke: Film Formation in Modern Paint Systems

ISBN 978-3-86630-861-9

numerator and denominator of the fraction are equal, cos $\alpha = 1$ and $\alpha = 0°$. In that event, wetting is just complete. Smaller, i.e., negative, values of α are not physically possible. In other words, even in the case of better wetting than in the extreme case above, α is always equal to zero, and Young's equation loses its validity. To cover this situation, we define a further constant called the **spreading constant**:

$$S = \sigma_s - (\sigma_{s,l} + \sigma_l) = \sigma_s - \sigma_{s,l} - \sigma_l.$$

As can be easily verified, S is just equal to zero in the extreme case described above and greater than zero when the fraction in Equation 3.2 has a value greater than one. The latter, however, means that the outward force at the edge of the drop (i.e. away from the centre) is always greater than the sum of the opposing forces. There is no equilibrium, and the droplet spreads out further and further in an increasingly thinner film, i.e. it **spreads** until it evaporates from the edge or is stopped by the "roughness mountains" of the subsurface.

To summarise:

- For a (liquid) coating material to yield a uniformly thick film and a strongly adhesive coating, it must wet the substrate well and spread over it as far as possible. This is achieved when the surface energy (surface tension) of the substrate is at least as large as the surface tension of the coating material ($\sigma_{s,l}$, in this case, is very small or almost zero [3, 7]).

Since the surface energy of solids is not measured easily or accurately, it is standard practice instead to use the **critical surface tension** as a way of measuring the surface energy of the substrate; this can be quickly and easily determined on a series of test liquids of known surface tension [13, 26]. These are subjected to simple wetting tests (dripping, brushing) to determine that one which spreads out flat over the substrate ($\alpha \approx 0°$, cos $\alpha \approx 1$). In that case, Equation 3.1 assumes a special form

$$\sigma_s \approx \sigma_{s,l/cr} + \sigma_{l/cr} \quad \text{(cr = critical)},$$

and since $\sigma_{s,l/cr} \approx 0$, then $\sigma_s \approx \sigma_{l/cr} = \sigma\chi_r$.

The critical surface tension σ_{cr} of the substrate is therefore equal to the surface tension of a liquid which only just spreads out over the surface. All liquids (paints, etc.) that have a lower surface tension than σ_{cr} wet the substrate completely.

Table 3.1 lists the surface tension values of various solvents, substrate materials and binders.

From Table 3.1, the following generalisations may be made. The surface tension of

- organic solvents is low to moderately high
- water is very high
- binders is moderately high (usually higher than that of solvents)
- substrates is low (PTFE, polyolefins PE, PP) to moderately high (depending on purity)

Surface impurities having a low surface tension, such as grime, hand sweat, certain types of dust particles, and processing auxiliaries will generally impair the wettability (and subsequent paint adhesion), so that, in extreme cases, partial or complete **dewetting** can occur after application. When this happens, the paint film develops thin patches, holes, craters, or retreats over a large area to form islands (see Figure 3.2).

Substrate wetting can basically be improved in two ways:

- raising or equalising the surface tension of the substrate by
  - cleaning, degreasing
  - targeted roughening (sanding, blasting)

Table 3.1: Surface tension values of selected solvents, substrates and binders

Liquid[1]	σ [mN m⁻¹]
hexamethyldisiloxane (silicone)	16
n-octane	21
2-propanol	22
acetone	24
butyl acetate	26
toluene	29
butyl glycol	30
dioxane	37
water	73

Substrate[2]	σ [mN m⁻¹]
PTFE	20
polypropylene	28 to 35
polyethylene	30 to 36
polycarbonate	35 to 37
polyester	40 to 45
aluminium (untreated)	33 to 35[4]
steel (phosphated)	43 to 46[4]
steel (untreated)	29[4]
glass	73[4]

binder[3]	σ [mN m⁻¹]
poly(meth)acrylates	32 to 41
alkyd resins	33 to 60
polyvinyl acetate	36
polyvinyl alcohol	37
melamine resins	42 to 58
epoxy resins	45 to 60
chlorinated rubber	57

[1] *From [20, 27]: slightly different values*
[2] *From [3, 7, 13]: some values significantly different*
[3] *From [3]*
[4] *Critical surface tension*

   - chemical pretreatment (oxidation of polyolefins, conversion of metals, e.g. phosphating or chromating)
- lowering the surface tension of the coating material with **additives which wet the substrate** (special surfactants or substances with surface-active properties)

The additives can lower the surface tension of a coating to about 20 mN m^{-1}, but this too can have negative side effects, such as foaming or **poor interlayer adhesion** (when over-painted). Such additives must there be chosen and admixed very carefully [27, 28].

Waterborne coatings (waterborne paints, etc.) naturally tend to have a higher surface tension than solventborne products. But they, too, can wet even nonpolar (pretreated) plastics, such as polypropylene, if the appropriate additives are used.

Poor substrate wetting can also be more or less hidden by generating a higher layer thickness and/or by a higher viscosity. A rapid increase in viscosity after application,

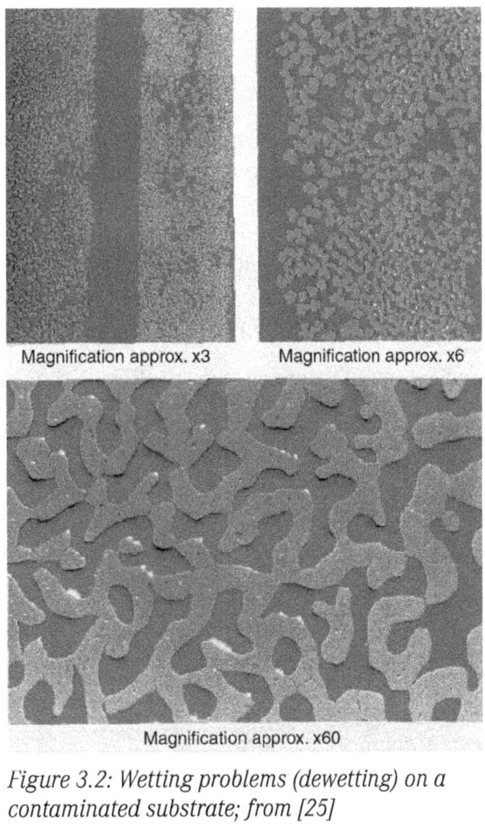

Figure 3.2: Wetting problems (dewetting) on a contaminated substrate; from [25]

due to a rapid drop in thixotropic shear thinning and/or fast surface drying, can set in faster than dewetting. The dangers here are that the drop in viscosity during subsequent heating causes the paint to retreat afterwards and that, despite looking as if it has formed a satisfactory film, the paint adhesion is patchy.

3.2 Levelling

Only the levelling of liquid coating films will be covered in this section. For levelling of coating powders, see Chapter 9.3.

All methods produce a wet film whose surface is characterised by a certain degree of surface unevenness (texture). Brush marks, tramlines and waviness caused by spraying are examples of this. Generally, the wet film should be self-smoothing and yield a flat surface when dry. An example of this smoothing process is that which is exhibited by the "**droplet mountain range**" which is present just after spraying (see Figure 3.3). The droplets must first flow together or **coalesce.** At this stage, the surface is still very uneven with "mountains" and "valleys", and that is when actual levelling starts.

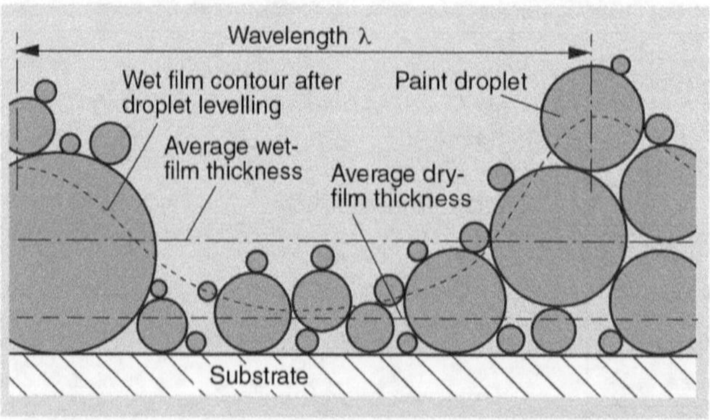

Figure 3.3: "Droplet mountain range" after spray application and immediately before coalescence and levelling; from [3]

The driving force behind levelling is the tendency of the film to lower its surface energy by decreasing its surface area, much in the way a bed sheet straightens out when it is pulled at the edges or corners. It can therefore be expected that the higher the surface tension of a paint film, the faster it will undergo levelling. Levelling, however, also entails material flow from the mountains into the valleys, and this proceeds slowly if the viscosity is high. Good levelling is therefore contingent on the viscosity's not rising too quickly after application. The reasons for an increase in viscosity may be:

- solvent/water evaporation (physical surface drying)
- molecular enlargement (incipient curing)
- decrease in the thixotropic shear thinning that occurred during application

The fundamental physico-mathematical description of the primary levelling process is provided by *Orchard* [13, 29]. He proposes a model of an idealized cross-section that has a sine wave profile (see Figure 3.4), and arrives at the following equation (**Orchard equation**)

Equation 3.3:

$$\ln\frac{a_0}{a_t}=\frac{16\pi^4\sigma x^3}{3\lambda^4}\int_0^t\frac{1}{\eta(t)}dt$$

a_0 is the starting amplitude (t = 0)
a_t is the amplitude after time t
x is the mean film thickness
σ is the surface tension of the film
λ is the wavelength of the sine curve (constant)
η(t) is the viscosity (time-dependent)

The integral is called the **fluidity integral**. If the viscosity remains more or less constant, it can be replaced by the product, η t (The precise mathematical derivation is very complicated. A simplified derivation, which, with the exception of the factor ($16\pi^4/3$), also leads to Orchard's equation, is given in Annex 1.)

From Equation 3.3, we can infer the following:

- levelling becomes progressively slower, and is
- accelerated moderately by a rise in surface tension, and
- is retarded by a rise in viscosity
- an increase in film thickness has a highly positive effect, while
- an increase in wavelength has a highly negative effect on levelling.

The last-mentioned means that short waves level faster than long ones.

In the literature, it is variously reported that gravity is partly responsible for levelling. Rough calculations, however, show that the gravity pressure exerted by a "paint mountain" of height a ($p = g\ \rho\ a$) is practically negligible compared with the excess pressure ($p = 2\ \sigma/r$) caused by the radius of curvature r. ("While a person unacquainted with the field might first think that levelling results from gravitational effects, this is clearly not, to a significant degree, the case" [22]).

A common cause of unsatisfactory levelling is an excessively rapid rise in viscosity. Although viscosity, as shown in Equation 3.3, is only of moderate importance mathematically, it often (for the reasons already described) undergoes quite a marked time-dependent change before film-formation occurs - not infrequently in the order of a power of ten or more - and it is therefore of considerable influence in practice.

Since low viscosity promotes levelling, but also unfortunately leads to undesirable sagging on vertical surfaces (see Chapter 3.3), the goal should not be low viscosity of itself, but rather an optimal viscosity-time profile immediately after application. In practical terms, this means that

- the volatility of the solvents,
- the temperature control and the air flow in the dry zone (evaporation zone, oven),
- the chemical reactivity of the coating (if of the curing type),
- and the type and amount of rheological additives (if any),

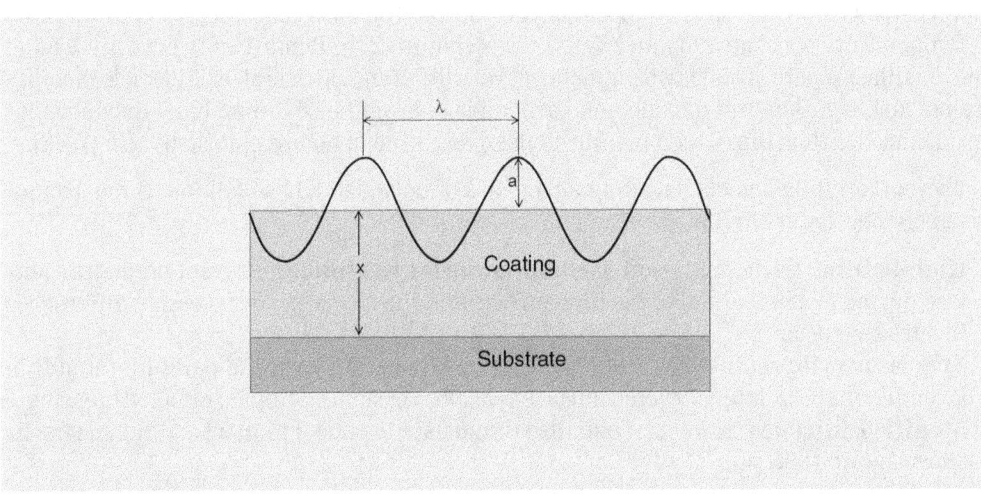

Figure 3.4: Profile model for calculating paint levelling according to Orchard; from [13]

must be coordinated – and that requires constructive collaboration between paint developer and paint processor.

A further cause of unsatisfactory levelling can be the existence of a yield point. This occurs when the liquid more or less stops flowing below the shear stress (see Chapter 1.4.2). As a result no further levelling takes place once a certain amount has occurred. Yield points occur with, e.g., dispersion-based waterborne paints, including clearcoats.

Now, it might be thought that optimising the parameters in Equation 3.3 or observing the above conditions would be bound to yield good levelling or a smooth paint. Unfortunately, that is not the case. While the surface tension is the driving force for levelling, inevitable differences in surface tension during evaporation lead to the formation of structures (texture), even if the film for a while was almost flat.

First, it should be noted that

- the surface tension rises as the temperature drops and as the binder concentration increases due to solvent evaporation
  (when water evaporates, the surface tension usually falls)

Since the evaporation rate and the associated concentrating and cooling processes (evaporative cooling) vary across the surface, differences arise in the surface tension of the film surface.

Now, it is a fact that:

- where a difference in surface tension occurs, material flows from the area of low tension to the area of high tension.

This leads to the formation of structures (unevenness), which can change both locally and temporally, until a viscosity increase or other factor causes it to cease. If, for example, incipient crosslinking or gelation prevent the structures from flowing together before stoving occurs, the outcome will be an uneven finish. A typical example of uneven spray coating is **orange peel**, shown in Figure 3.5. It can occur at wavelengths of 0.2 to 10 mm and is usually measured directly on the finished painted object with the aid of hand-held laser devices [30].

A special type of structure formation encountered occasionally is that of **Bénard cells**. These are currents of paint material circulating between the substrate and the paint surface that mostly assume the shape of hexagonal columns. They are caused by local differences in temperature, concentration and density (see Figure 3.6). Bénard cells not only lead to uneven films. Where at least two pigments of very different particle size or particle mobility are present, e.g. titanium dioxide and carbon black or phthalocyanine blue, they also are responsible for **floating**, which manifests itself as a kind of honeycomb structure [7, 13].

To prevent levelling defects, levelling agents are often added. These fall into three groups, based on their nature and mode of action:

- **High-boiling; with very good solvent power for the binder**: Prevent premature surface drying (solidification) of the film surface and the formation of excessive differences in surface tension.
- **Special acrylic resins** (e.g., butyl acrylate copolymers): Cover the paint film in a thin layer of equal but barely lowered surface tension. Possible additional effect: degassing
- **Modified silicones**: Lower and equalise the surface tension. Frequent additional effects: substrate wetting, slip.

3.3 Sagging

Sagging of a wet film or melt on a sloping or vertical surface is not part of film formation, but it can disrupt or prevent the formation of a smooth film.

A liquid film will always flow more or less rapidly down a vertical or inclined surface unless it is prevented from doing so by a sufficiently high yield point.

For paint technologists, the term sagging does not refer to this primary flow process, but rather the subsequent consequent formation of **runs**. These are subdivided in accordance with their appearance into **curtains** (wide) and **tears** (narrow), see Figure 3.7.

Were the paint film to flow over the entire substrate surface at the same rate, there would be no runs on the surface. However, since the flow rate increases markedly with increasing layer thickness in accordance with

Equation 3.4: $$v_{max} = \frac{g\rho x^2 \sin\alpha}{2\eta}$$

v_{max} is the flow rate of the film surface
g is the acceleration due to gravity
ρ is the paint density
α is the inclination to the horizontal
η is the dynamic viscosity
x is the film thickness

the inevitable film unevenness - along with local viscosity differences - always lead to unequal flow rates and thus to local material accumulation, which then evolve into runs.

The maximum shear stress in a paint film is $\tau max = g\,\rho\, x \sin\alpha$. If the layer thickness x is so small that τmax is below the yield point (the minimum shear stress for flow), no flow or sagging will occur at all. This case is also covered by Equation 3.4, insofar as, at a yield point, the viscosity tends to infinity and the flow rate thus tends to zero.

Flow of the film is retarded by the viscosity, and so the increase in viscosity delays the formation of runners. If drying is fast enough to increase the viscosity, it is not

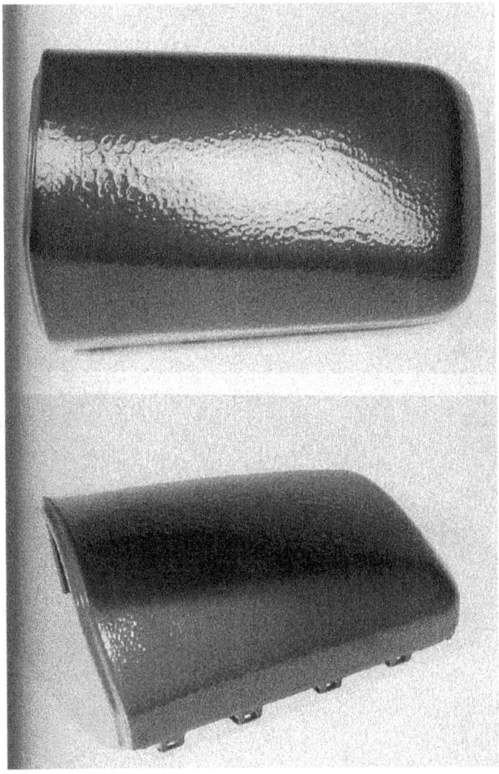

Figure 3.5: Poorly levelled spray paint (orange peel); from [25]

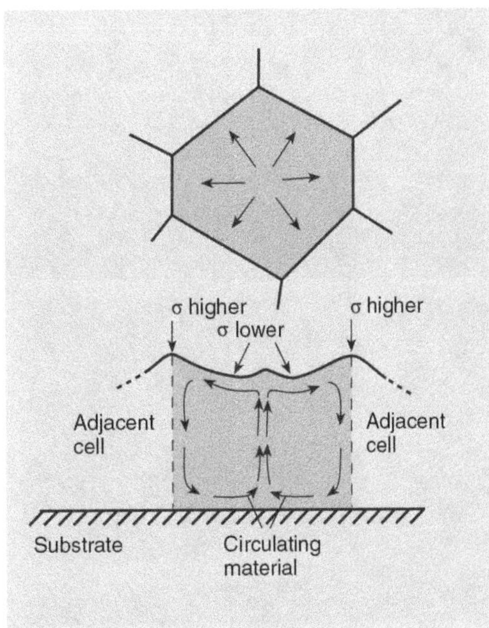

Figure 3.6: Formation and shape of a Bénard cell; cross-section (bottom) and plan view (top)

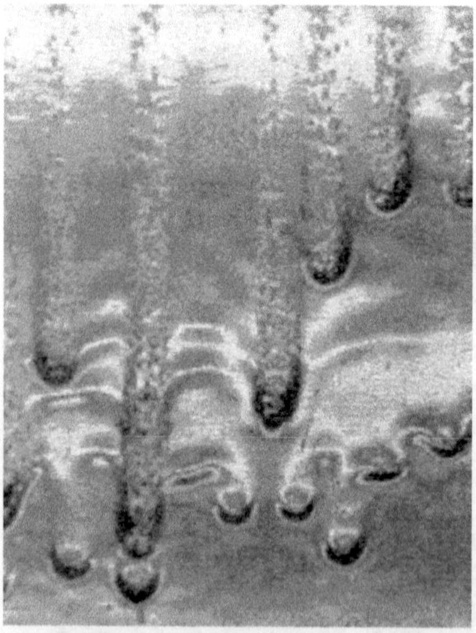

Figure 3.7: Sagging of paint in the form of tears and curtains; from [25]

possible for visible sagging to develop in the first place. However, where this is not the case, which is especially true for low-solvent, high-solid coatings, thickening agents or, preferably, thixotropic agents, also known as anti-sagging agents, need to be added to suppress sagging. With proper selection and dosage, the latter ensure that a low-viscosity coating will have finished levelling before sagging is prevented by any increase in viscosity.

Generally, the tendency to form runs increases with increase in layer thickness, not just for purely mechanical reasons (in accordance with Equation 3.4), but also because thickening due to physical drying slows down as the layer becomes thicker (see Chapter 6.1.2.2). The maximum thickness of a layer of wet paint on a vertical surface at which sagging does not occur is the **sagging limit**. This can be determined relatively easily with a **levelling test blade**.

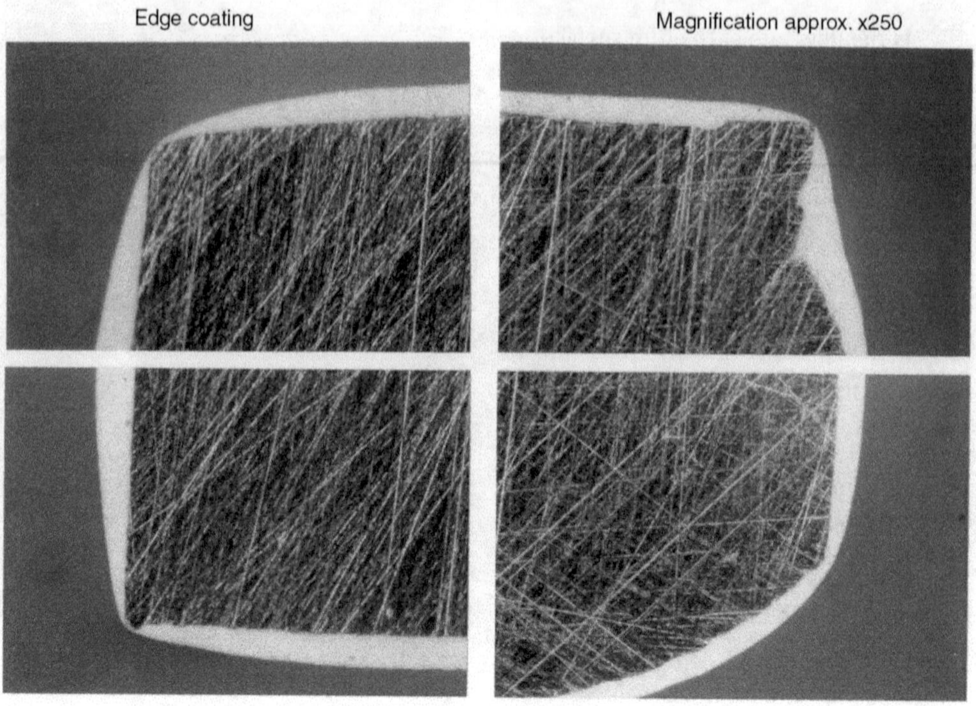

Figure 3.8: Transverse sectional view of a paint finish exhibiting edge pulling (magnification: x250); from [25]

3.4 Edge pulling

A paint film covering an edge (a cutting edge, stamping edge, etc.) has a curved surface which is under a higher pressure, relative to the flat film, given by $p = \sigma/r$ (where r is the radius of curvature and σ is the surface tension of the film). This excess pressure forces the paint material away from the edge into the adjacent film areas, where temporary beading occurs. The result of this edge pulling is that the layer over the edge is too thin or almost completely absent and thus offers little protection against mechanical and corrosive agents (see Figure 3.8).

Ways to avoid edge pulling include lowering the surface tension and/or raising the viscosity, the latter usually being more successful. Since edge pulling of the wet film ceases after a few seconds [26], the viscosity would need to be high enough straight away, but that would impair levelling. For this reason, edge pulling can never be avoided completely. In isolated cases, it can be advantageous to apply an extra coating to the edges during electrostatic and electrodeposition [17]. However, if the paint turns very mobile in the oven, this initially positive effect will be all but lost again. For a fundamental, physical treatment of edge pulling, see [13].

4 Physical principles of paint drying and curing

4.1 Convection drying

Convection drying – placing the painted object in a current of hot air or blowing hot air onto it – is the classical and still the most common technique for drying paint at elevated temperatures. The object to be painted is placed in a convection oven (hot-air oven, forced-circulation oven) – either at rest or undergoing slow translation. The air is mostly fed in a loop and is continuously enriched with only as much fresh air as is needed to prevent explosive atmospheres from being formed with the organic compounds evaporating from the layer or the wet film. The upper limit for the concentration of organic vapours in the oven air is, as per the German regulation BGV D25, 0.8 vol.%.

The heat is transferred by convection from the dryer air into the film and hence indirectly to the object's surface. This is usually taken to mean convective heat transfer from a flowing fluid to a contiguous surface and vice versa. Convective heat transfer may be described by the following empirical equation:

Equation 4.1:

$$\dot{Q} = \alpha A (\vartheta_A - \vartheta_F) = \alpha A \Delta\vartheta = \alpha A \Delta T$$

$\dot{Q}$ is the heat flow across surface A
α is the heat transfer coefficient
ϑ_A is the air temperature in Celsius
ϑ_F is the temperature of the film/layer surface in degrees Celsius
T is the temperature T (absolute temperature) in Kelvin
ΔT is the temperature difference[9)]

When a fluid flows along a surface, its speed decreases from the full value of vL (at some distance away from the surface) to zero at the surface on account of the drag exerted by its viscosity within the fluid boundary layer of thickness δ (see Figure 4.1). If we assume that the heat transfer is a special heat-conduction process through the fluid boundary layer of mean thermal conductivity, λBL, the equation above can be written:

$$\dot{Q} = \lambda_{BL} \frac{1}{\delta} A \Delta T, \text{ also } \alpha = \frac{\lambda_{BL}}{\delta}$$

Since the **boundary layer thickness** decreases with increase in fluid velocity, the heat flow rises with increase in velocity of the dryer air. This is generally between 2 and 5 m s^{-1}, but on direct blowing can be as much as 15 m s^{-1} [7, 15]. The points to remember are that: Heat flow into the film (and thus the object) increases with

- increase in the speed of the dryer air around the film or the object and with
- increase in the difference between the air temperature and the temperature of the film surface.

[9)] Temperature differences in industry should be quoted in Kelvin (K). They are numerically equal to the differences in °C.

Peter Mischke: Film Formation in Modern Paint Systems

ISBN 978-3-86630-861-9

The temperature in the film is all-important, because the film will only dry or cure sufficiently, if it has been maintained long enough at a sufficiently high temperature. The values for temperature, time and any further technical conditions, such as air humidity for optimum and complete film formation, are collectively known as the **stoving** or **drying conditions;** the tolerance ranges for these variables constitute the **stoving window.** If the minimum stoving time and/or temperature is not reached, **under-stoving** occurs; its opposite is **over-stoving**. For practical purposes, it is extremely important to distinguish between the object temperature and the circulating-air temperature, and between the duration (**holding time**) of a particular object or stoving temperature and the **oven dwell time.**

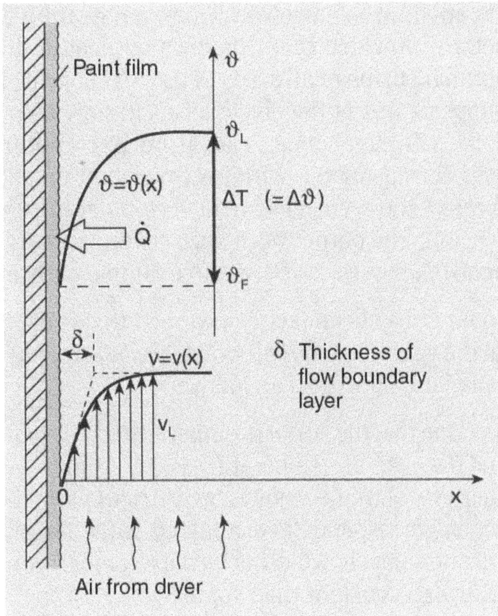

Figure 4.1: Fluid boundary layer and temperature profile during heat transfer (see text for explanation)

In a very simplified model, assume that air at a temperature of ϑ_A = 150 °C flows identically around the upper and lower surfaces of two uncoated steel plates of thicknesses d_1 = 7.5 mm and d_2 = 0.75 mm, which are at an initial temperature ϑ_0 = 25 °C, starting at time t = 0. Let the heat transfer coefficient, as per the rough equation $\alpha/\text{W m}^{-2}\text{K}^{-1} = 5.6 + 4\, v_L/\text{m s}^{-1}$ [32], be 20 W m^{-2} K^{-1}, the density of the steel be $\rho = 7.8 \cdot 10^3$ kg m^{-3}, and its specific heat capacity be C = 0.50 kJ kg^{-1} K^{-1}. The question is: how quickly would these plates heat up to a given temperature ϑ? The heat balance is given by:

$$CV\rho\, d\vartheta = \alpha A(\vartheta_A - \vartheta)dt$$

where A is the area cross-section, V is the plate volume = A · d,

Integration:

$$\int_{\vartheta_0}^{\vartheta} \frac{d\vartheta}{\vartheta_A - \vartheta} = \frac{\alpha 2A}{CV\rho}\int_0^t dt$$

leads to the equation:

$$\vartheta = \vartheta_A - (\vartheta_A - \vartheta_0)e^{-\frac{2\alpha}{C\rho d}t}$$

For flow to one side of the plate, factor 2 does not apply.

Substituting the above data and different heating times t (oven dwell times), we arrive at the the following sample results:

- plate with d_1 = 7.5 mm: 100 °C after 10 min, 139 °C after 30 min
- plate with d_2 = 0.75 mm: 100 °C after 1 min, 139 °C after 3 min, 148 °C after 5 min.

So, if a stoving temperature (object temperature) of 150 °C is required, then the thin plate needs approx. five minutes' heating time for double-sided flow, i.e. 5 minutes must be added to the oven dwell time; for single-sided flow, the heating time would be approx. 10 minutes.

In general, then, there is a risk that the thick sections of an object might remain cold during the period in the oven if the air flow around them in the oven is not favourable, i.e. if they are shielded in some way. However, heating can be accelerated somewhat by direct jets of hot air, provided that this does not lead to film-forming problems, such as skinning. For optimal curing of all areas of painted objects, it is therefore necessary to record the surface temperatures of the object as accurately as possible as it passes through the oven. This is done with non-contact **radiation pyrometers** and – additionally for the purpose of process optimisation – with heat-resistant data loggers, which accompany the original object through the oven, measuring its temperature profile as a function of time via numerous heat sensors. The output from such "oven trackers" can be displayed as an **oven temperature profile**. Figure 4.2 shows an example of three such profiles.

So far, it has been tacitly assumed that the temperature of the film is roughly equal to that on the surface of the substrate (the object temperature). But is this true? Here, again, a quick calculation provides an answer:

Assume that the surface temperature of the substrate during the heating phase is $\vartheta S = 130\,°C$ and the oven-air temperature is $\vartheta A = 150\,°C$. Again, let $\alpha = 20\ W\,m^{-2}\,K^{-1}$. Let the thermal conductivity of the (already largely dried) paint film of thickness $x = 40\ \mu m$ be $\lambda = 0.2\ W\,m^{-1}\,K^{-1}$, which corresponds to a rounded value for plastics and other thermal insulators [32]. The question now is what the surface temperature ϑF of the film would be in the steady state, i.e. over a constant time span. So,

$$\dot{Q} = \frac{\lambda}{x} A(\vartheta_F - \vartheta_S) = \alpha A(\vartheta_A - \vartheta_F)$$

Solving for ϑ_F, we get $\vartheta_F = 130.08\ °C$. In other words, in convection drying, the film is practically as hot as the substrate across its entire thickness.

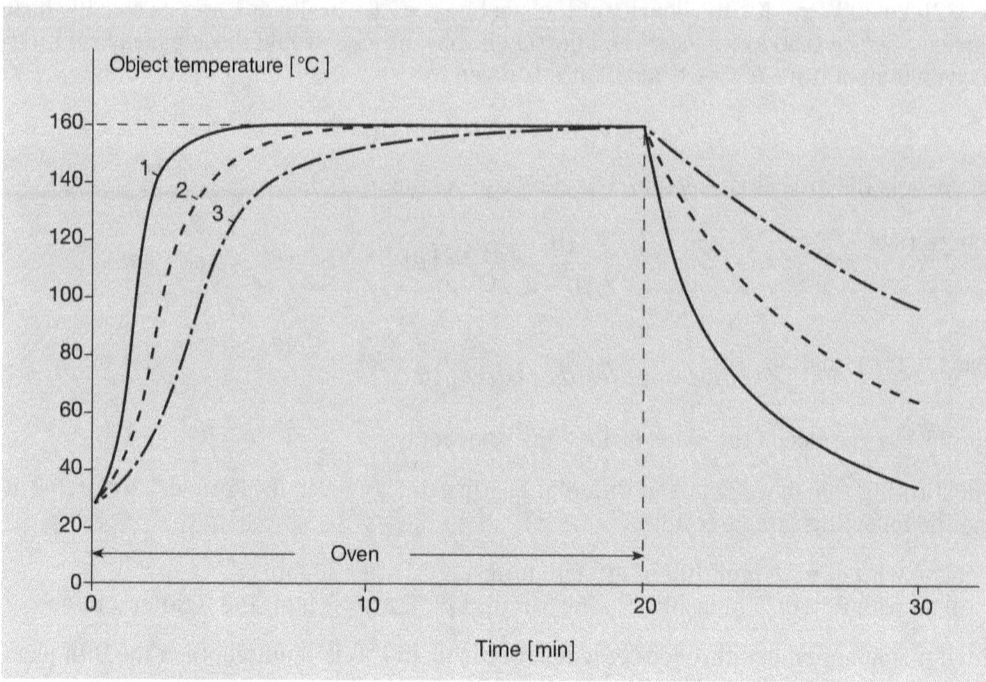

Figure 4.2: Oven temperature profiles
1 thin material; good air flow
2 thick material and/or poorer air flow; curing just about adequate
3 very thick material and/or defective air flow; curing inadequate

Even the surface of a loose layer of coating powder 150 µm thick in the 150 °C oven air would have only a slightly higher temperature (at most 132 °C) than the surface of the 130 °C substrate.

Despite these restrictions on the film temperature, over-stoving can occur, even in convection drying, if

- the oven air is at the desired stoving temperature, but the dwell time in the tunnel oven increases because the conveyor belt stops
- or - much worse - the oven-air temperature, for the sake of a very short dwell time and thus rapid substrate heating, is much higher than the maximum stoving temperature, but then a long dwell time occurs.

The conditions in the latter case are precisely those which obtain in **coil coating.** The dwell time here ranges from 10 s to 2 min because of the high belt speed of up to 250 m min^{-1}. In order that the belt which is bearing the film for curing may be heated sufficiently (peak temperature: 200 to 270 °C), the oven air must be much hotter - as high as 400 °C [7, 22].

The total quantity of heat Q_{tot} flowing into the object to be coated is the aggregate value of the heat requirements for heating the object (Q_O), heating the film (Q_F) and the evaporation of volatile components (Q_V). A curing reaction provides some heat (Q_R).

$$Q_{tot} = Q_O + Q_F + Q_V - Q_R$$

Although not of great practical relevance, let us estimate the approximate heat quantities involved. For this, consider a clearcoat film, composed additively of a water layer 50 µm thick and a layer of reactive resin (polyol/polyisocyanate) 50 µm thick, on a steel sheet 0.75 mm thick. Further, let the temperature be raised from 25 to 150 °C. Result (rounded values):

Heating of steel sheet (0.75 mm):	Q_O = 360 kJ m^{-2}
Heating of film (water content up to 100 °C):	Q_F = 28 kJ m^{-2}
Water evaporation:	Q_V = 110 kJ m^{-2}
Heat of chemical reaction ($-\Delta_R H$):	Q_R = 10 kJ m^{-2}

With regard to the overall heat balance for the drying process, so much heat is required for heating the fresh air, for compensating the heat loss through the oven openings and walls and for heating the goods carrier and perhaps solid objects that the level of heat consumed by the film is virtually irrelevant [7].

4.2 Heat transfer by means of infrared radiation

4.2.1 Physical principles

Like light, infrared radiation (IR radiation), which is also called ultra-red or thermal radiation, is electromagnetic radiation. The various types of electromagnetic radiation are shown in Figure 4.3. As may be seen, IR radiation is a direct continuation of the red (long-wavelength) side of the visible light spectrum. The entire wavelength range extends from 0.78 µm to 1 mm. According to the equation $E = h\nu = h\,c / \lambda$ (where h is Planck's constant ($6.626 \cdot 10^{-34}$ J s), c is the speed of light ($2.998 \cdot 10^{8}$ m s^{-1}), ν is the frequency, and λ is the wavelength), the maximum energy of the IR photons is 153 kJ mol^{-1}, equivalent to 1.59 eV (electron volts). This energy is less than the bond dissociation energy of almost all chemical bonds. IR radiation is not capable of breaking stable chemical bonds direct. However, it

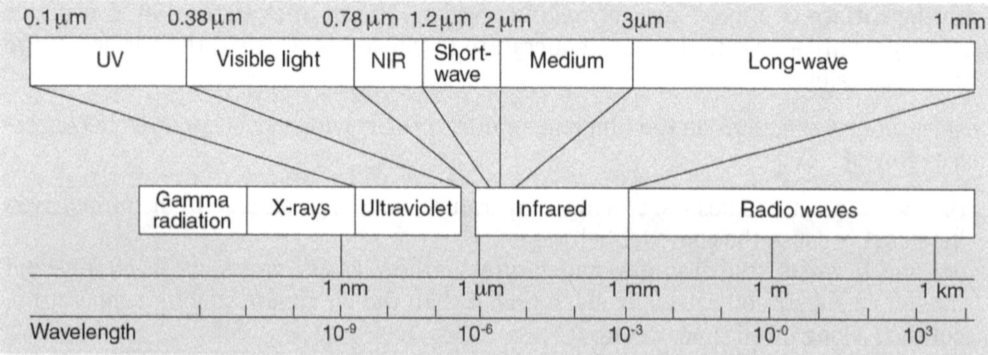

Figure 4.3: Classification of electromagnetic radiation by wavelength range; from [7]

causes them to vibrate, which ultimately manifests itself as heat, which in turn can lead to chemical reactions and bond breaking.

Whereas matter becomes warm when it absorbs IR radiation, hot bodies will emit IR radiation. Any body of finite temperature (above zero Kelvin) emits infrared radiation – in this case it is specifically called **thermal radiation**. According to Planck's law of spectral radiance of electromagnetic radiation (see physics textbooks), radiant power is distributed across the wavelengths as a function of temperature in a characteristic way (see Figure 4.4)[10)].

As may be seen, the radiation is displaced to shorter wavelengths as the temperature rises. This shift is described by **Wien's displacement law:**

Equation 4.2:

$$\lambda_{max} = \frac{2897.8\,K\,\mu m}{T}$$

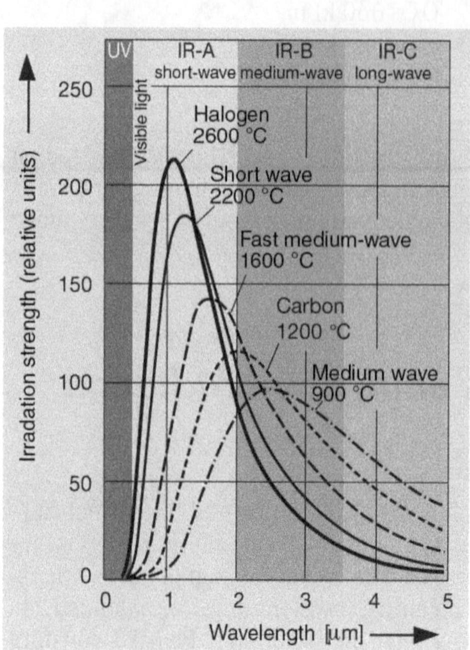

Figure 4.4: Spectral distribution of radiant power at various IR emitter temperatures; from [7]

$\lambda_{max.}$ is the maximum radiant power per wavelength

T is the absolute temperature (Kelvin temperature)

It may also be seen that, as the temperature rises, more and more of the radiation appears in the visible range. The hotter an object is, the stronger (brighter) it glows.

The areas under the curves represent the total power P emitted per unit area across all emitted wavelengths. P, too, can be described by a surprisingly simple equation, called the **Stefan-Boltzmann law (T to the fourth power law)**

Equation 4.3:

$$P = \varepsilon C_s \left(\frac{T}{100}\right)^4$$

P is the radiant power per unit area

C_S is Stefan's constant for a black body ($5.67\ W\ m^{-2}\ K^{-4}$)

[10)] Max Planck, 14 December 1900: Formulation of the "Quantum Hypothesis"

Table 4.1: Radiation data for IR emitters (literature data are highly inconsistent) [7, 11, 15]

Type of emitter	Surface temperature [°C]	Wavelength of the radiation maximum [µm]	Theoretical radiant power (ε =1) [kWm^{-2}]	Actual radiant power [kWm^{-2}]
long-wave	300 to 600	5.1 to 3.3	6.1 to 33	up to 20
medium-wave	600 to 1500	3.3 to 1.6	33 to 560	8 to 50
short-wave	1500 to 3000	1.6 to 0.9	560 to 6508	20 to 100
NIR	3300 to 3500	0.81 to 0.78	9242 to 11492	>100

ε is the emissivity of the body ($\varepsilon \leq 1$)
T is the absolute temperature (Kelvin temperature)

In other words, emission of radiation increases markedly as the temperature rises. Perfect black bodies (idealized) radiate the most strongly. All real bodies are more or less "grey" and radiate to a lesser extent. This is expressed by the emissivity, ε, which indicates the ratio of the energy radiated by a grey body to energy radiated by a black one.

4.2.2 Industrial infrared lamps

Industrial IR lamps are generally thermal emitters, i.e. they contain hot, radiant surfaces or bodies. However, **IR lasers** are used too. Heating and cooling times vary with design, but fast responses are sought for the purpose of close control [3, 11, 15]. Since an IR dryer does not dry via a tangible medium (air), whose temperature can be set and measured directly, it is even more important to use a radiation pyrometer to measure the object's surface temperature in IR drying than it is in convection drying.

Emitters are generally – and, in the literature, not consistently – classified by their peak wavelengths into long (λ_{max} = 5.1 to 3.3 µm), medium (λ_{max} = 3.3 to 1.6 µm), short (λ_{max} = 1.6 to 0.9 µm) and very short-wavelength types (NIR emitters, λ_{max} around 0.8 µm). Table 4.1 gives an overview of the radiation data. Long-wavelength emitters whose temperature is below 700 °C and therefore do not emit visible incandescent light are also called dull emitters. Hotter emitters which emit shorter wavelengths are generally described as bright emitters.

4.2.3 Radiation input into the object for coating

When two large flat surfaces of different temperature are parallel to each other, the Stefan-Boltzmann law (Equation 4.3) states that the warmer surface radiates at a higher intensity towards the colder one; while the colder one radiates at a lesser intensity towards the warmer. The effective heat flux transferred (corresponding to the radiant power) is

Equation 4.4:

$$\dot{Q} = A\,C_{1,2}\left[\left(\frac{T_1}{100}\right)^4 - \left(\frac{T_2}{100}\right)^4\right]$$

with

$$C_{1,2} = \frac{1}{\frac{1}{\varepsilon_1} + \frac{1}{\varepsilon_2} - 1} C_s$$

$\dot{Q}$ is the transferred heat flux
A is the area cross-section (the same for both surfaces)
$C_{1,2}$ is the mean Stefan constant

T_1, T_2 are the absolute temperatures of the warm/cold surfaces
$\varepsilon_1, \varepsilon_2$ are the emissivity values of the surfaces
C_s is the Stefan constant for a black body (5.67 W m^{-2} K^{-4})

It is immediately apparent that there is no effective heat flux when $T_1 = T_2$, which is plausible and also a necessary condition of the second law of thermodynamics. The conclusion to be drawn is that heat transfer depends not only on the temperature of the emitter (transmitter) but also that of the object (receiver).

In general:

- IR radiation can shorten heating times much more than hot air can.

If the irradiated surface or cross-section thereof is inclined at an angle, α, to the emitter surface (i.e. $\alpha = 0$ in the case of parallelism), the transferred heat flux is reduced by the factor cos α. Specific formulas exist for calculating radiation exchange in geometrically complex cases [35].

Film formation requires that a somewhat more nuanced view be taken of the processes occurring on the surface covered with the paint. When IR radiation impinges on the paint film, it is split into at most three portions: α, the absorbed portion which is transformed into heat; τ, the transmitted portion; ρ, the reflected portion.

All three portions are wavelength dependent. Generally, for each wavelength

$$\alpha(\lambda) + \tau(\lambda) + \rho(\lambda) = 1$$

α is the degree of absorption
τ is the degree of transmission
ρ is the degree of reflection.

Furthermore, a thought experiment can show that $\alpha = \varepsilon$ in all cases.

In other words, substances that are good absorbers are also good emitters. Substances that are good reflectors will absorb and emit little radiation. Here are some ε values [32, 36]:

aluminium (polished): 0.04 to 0.10; iron (matt): 0.25 0.4; steel (rolled): 0.67; paints: 0.80 to 0.97; wood: 0.85 to 0.95, glass, ceramics: 0.90 to 0.95.

The extent to which paint films absorb IR radiation varies with their composition and the wavelength or wavelength range of the radiation. The proportion of radiation absorbed primarily in the paint also depends of course on the layer thickness, because the radiation is not swallowed directly at the surface, but rather penetrates into the film, undergoing exponential attenuation as it penetrates further.

Equation 4.5:

$$I(x) = I_0 \, e^{-Ax}$$

$I(x)$ is the radiant intensity at depth x in the film
I_0 is the incident radiant intensity
A is the absorption coefficient (wavelength-dependent)

A, for example, can be expressed in the form $A = 0.693 / x_{1/2}$, where $x_{1/2}$ is the "half-life depth", i.e. the depth in the film (from the surface) at which the initial intensity I_0 has dropped to $^1/_2\, I_0$. Radiation which penetrates as far as the substrate is partially reflected by it back into the film and partially transformed into heat. A special case is presented, for example, by metallic paints in which the aluminium flakes reflect a significant proportion of the radiation. Objects coated with such paints heat up much more slowly in infrared heating zones than uni-coated ones (see Figure 4.5).

Generally, absorption of IR radiation by paints exhibits a marked spectral dependence. Most paint films display absorption behaviour similar to that outlined in Figure 4.6. This wavelength dependence of absorption leads to specific effects which will be discussed broadly below [6, 7, 11].

Dull emitters radiate in the region around 4 μm, where paints exhibit little absorption. The radiation therefore penetrates for the most part as far as the substrate, where it is absorbed or reflected to an extent depending on the radiation properties of the substrate material. The outcome is the temperature profile shown in Figure 4.7, i.e. the paint is primarily heated from the substrate out. The corollary of this is that the risk of **popping**, which is high in the case of thick wet films, is reduced and levelling is improved, because the surface of the film stays "open" (retains a low viscosity) for longer and skinning does not occur. Dull emitters, however, cause extensive convective heating of the oven air thanks to their large surface area, and so additional convection drying occurs from the surface down.

The transition to bright emitters takes in the wavelength range 3.5 to 2.5 μm. This is a region of strong CH, NH and OH absorption bands due to valence vibration. The upper regions of the paint film absorb the IR radiation, to yield a temperature profile as shown in Figure 4.7b. Heat input is high, but rapid surface drying in the case of thick films can cause problems with film formation.

The behaviour of the paint film at wavelengths below 2 μm is more dependent on the pigmentation. Generally, transmission decreases and the ratio of reflection to absorption increases as the pigmentation becomes lighter [6, 7]. Every reader will surely have learned from experience that cars painted black heat up faster in sunlight than do lighter or even silver coloured ones.

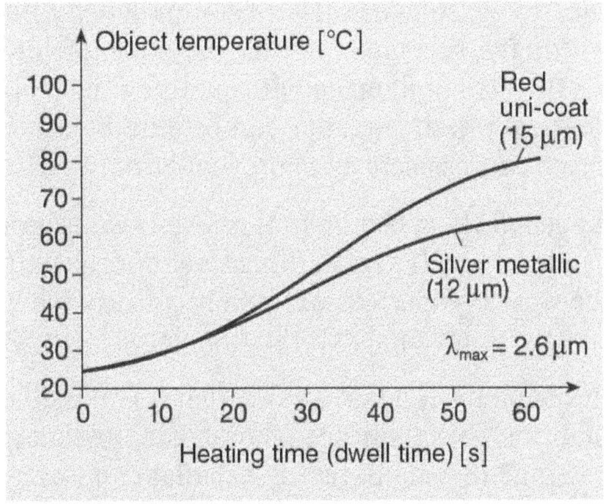

Figure 4.5: Heating rate under IR irradiation (radiation maximum at 2.6 μm) of a carbody section in the intermediate dryer as a function of pigmentation of the waterborne paint (source: BASF)

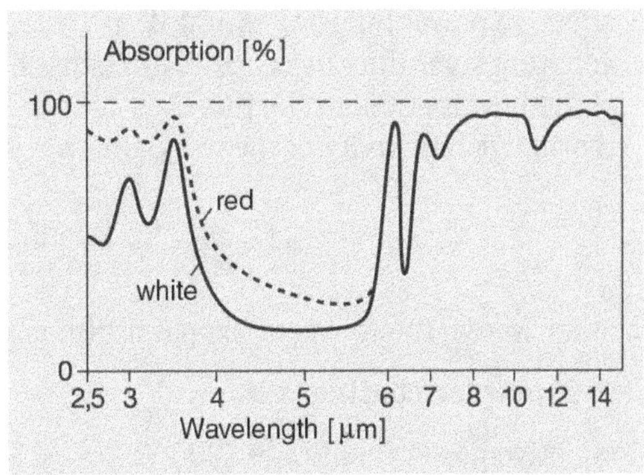

Figure 4.6: Typical IR absorption spectrum of paints (alkyd/melamine, white and red); spectra are virtually identical above 6 μm (adapted from [7])

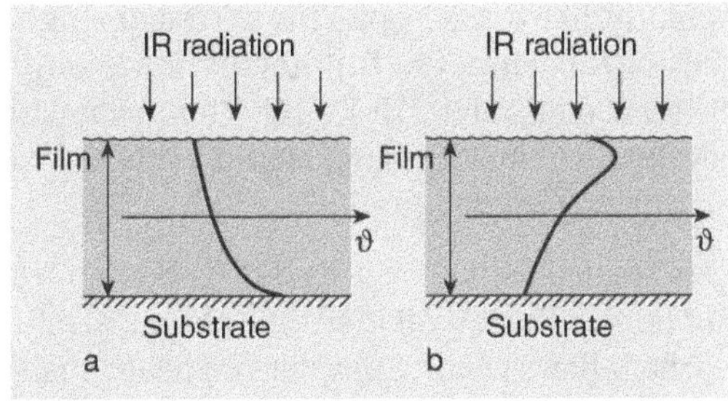

Figure 4.7: Temperature profile in a paint film of variable IR absorption (schematic), adapted from [11]

a) slight absorption
b) high apsorption

The very high power density of NIR emitters causes the paint layer to heat up extremely quickly. This property is mainly put to use for curing coating powders on heat-sensitive substrates, such as MDF (medium density fibre board). The layer of coating powder melts within five seconds, and then cures completely in a few seconds, without the substrate being over-heated or damaged during this time [16]. But wet films that have undergone sufficient physical pre-drying, too, can be cured with NIR radiation. This of course necessitates precision process control to prevent bursting or over-stoving of the film.

A special effect occurs in the case of waterborne paints. Water has a very strong absorption band at 3 μm. If a fresh film of waterborne paint is irradiated strongly with this wavelength, the water evaporates extremely quickly whereas any readily soluble **cosolvents** (organic auxiliary solvents) exit the film very slowly [37].

In Chapter 4.1, it was shown that it is possible for paint films not to become much warmer than the object surface when heated by hot air. But what happens in the case of infrared heating? Instead of trying calculate an exact answer – which is not possible with "pencil and paper" at any rate – let us use Equation 4.1 to estimate it. Again, let us assume that, within a short time space, there are hardly any changes in temperature (which is of course not true for very rapid heating).

Case 1: **wet paint**

- effective power input, $\dot{Q}_A$ = 50 kW m^{-2}
- apparent layer thickness x = 20 μm (smaller than paint film thickness because the focus of the heat input is in the film)
- thermal conductivity of the wet paint, λ = 0.2 W m^{-1} K^{-1}

$$\Delta T = \frac{\dot{Q}x}{A\lambda} = \frac{50 \cdot 10^3 \ W \cdot 20 \cdot 10^{-6} \ mmK}{1 \ m^2 \cdot 0.2 \ W} = 5 \ K$$

In other words, the inside of the paint film is approx. 5 K warmer than the substrate.

Case 2: **powder coating**

- effective power input $\dot{Q}_A$ = 100 kW m^{-2}
- apparent layer thickness x = 100 μm
- thermal conductivity of the powder layer λ = 0.03 W m^{-1} K^{-1} (air: 0.02)

Result: ΔT = 333 K.

Actually, however, despite the extremely high power density when a powder coating undergoes NIR curing, the film temperature only reaches about 230 °C, i.e. the temperature difference is some 210 K [16]. The explanation for this apparent discrepancy is that the powder layer melts during heating, and the sudden good thermal conductivity then greatly retards further heating.

Even if such estimates are a bit suspect to "theorists", they at least show that a paint layer can be significantly hotter than the substrate in the case of infrared radiation. This in fact is what allows coating powders on heat-sensitive substrates to be cured and accelerates the evaporation of water from waterborne paint films prior to the actual stoving process.

4.3 UV irradiation (for radiation curing)

4.3.1 General information on UV radiation

Ultraviolet (UV) radiation photochemically triggers polymerisation reactions that lead to crosslinking (curing). The reaction is largely complete after just a few seconds' exposure and the parts can be immediately processed, stacked or packaged. Because curing is fast, only short drying tunnels are required. A further advantage of radiation curing is that the coating materials release hardly any organic solvents because they are 100 % liquid or are emulsified in water. Since its introduction in the 1960s, UV technology has become very widespread and has evolved into the standard method in such industries as wood/furniture coating and the printing sector. Radiation curing is making increasing inroads into other areas, such as automotive OEM production line painting [11, 38, 39].

UV radiation is a continuation of the short-wavelength (violet) side of the visible light spectrum (see Figure 4.3). The wavelength regions of industrial relevance are divided up as follows:

- UV-A: 380 to 315 nm
- UV-B: 315 to 280 nm
- UV-C: 280 to 100 nm

Furthermore, a distinction is also made between quartz UV and vacuum UV. The former ranges from 380 to 200 nm, quartz is completely permeable to UV in this range. The vacuum UV range lies between 200 and 100 nm. Oxygen absorbs UV radiation at wavelengths below 200 nm (λ_{max} = 187 nm) and ozone forms. Consequently, UV irradiation in this region can only be performed in a vacuum.

Formation of **ozone** (O_3) cannot be entirely excluded in longer-wavelength industrial emitters either, since small proportions of radiation below 200 nm may be present. Ozone is toxic, has an unpleasant smell, and must therefore be extracted from the emitter chamber.

In accordance with the equation $E = h\nu = h c/\lambda$, photon energy increases over the wavelength range from 315 kJ mol^{-1} (3.26 eV) to 1196 kJ mol^{-1} (12.4 eV). Thus, UV radiation is capable of exciting bonding electrons in molecules and of breaking bonds. This fact is exploited by treating the coating material with a photo-initiator, which is cleaved in high yield by longer-wavelength UV radiation (UV-A), forming free-radicals or ion fragments that initiate the crosslinking reaction (see Chapter 10).

4.3.2 UV emitters

The UV emitters employed in radiation curing are variants of **mercury vapour lamps** and, to a lesser extent, xenon and other lamps.

Mercury vapour lamps consist essentially of some mercury contained in a quartz tube. Incandescent electrodes in the ends of the tube generate a gas discharge in the mercury vapour, and a line spectrum is emitted. The vapour pressure of the mercury determines whether the emitters are classified as **low-pressure**, **high-pressure** or **maximum-pressure lamps** [40]. The last of these are of no significance. Low-pressure lamps resemble common fluorescent lamps and are sometimes used for **pre-gelling** on account of their low radiant intensity [41]. Particularly noteworthy are **TL-03-low-pressure lamps**, because much of their radiation is not absorbed by titanium dioxide, a white pigment, and is useful for curing thick, pigmented paint layers [5].

High-pressure lamps, also called medium-pressure lamps, are the standard lamps, as it were, of UV curing. Their power consumption lies between 80 and 240 W per cm of tube length. Of this power (we are talking here of undoped lamps with electrodes or "**arc lamps**"), some 7 % is emitted in the UV-A range, 8 % in the UV-B, 10 to 15 % in the UV-C, 55 to 60 % in the infrared, with the remainder being visible light [42, 43]. Filters can be used to filter out short-wavelength radiation, e.g. to avoid ozone formation, but that may lead to impaired surface curing [43].

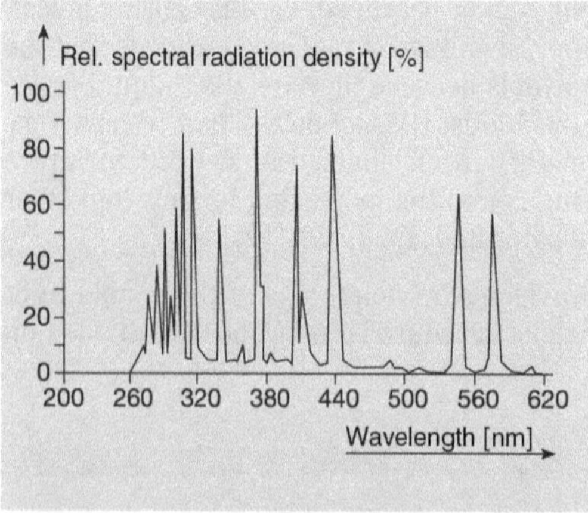

Figure 4.8: Emission spectrum of an undoped, high-pressure mercury vapour lamp

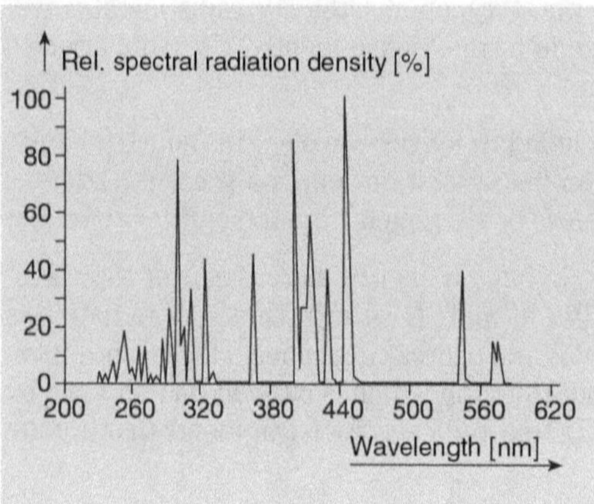

Figure 4.9: Emission spectrum of a gallium-doped, high-pressure mercury vapour lamp

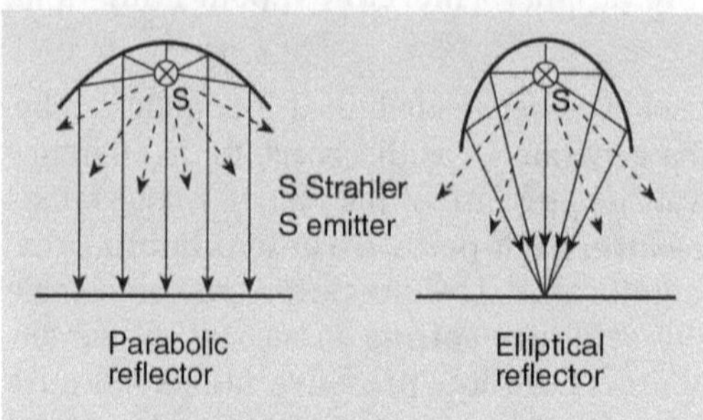

Figure 4.10: Radiation beams in parabolic and elliptical reflectors (simplified)

Figure 4.8 shows the spectrum of an undoped mercury lamp. It can be seen that a large proportion of the radiation is clearly in the UV region. Curing of pigmented UV coatings, however, requires lamps that are strong emitters at the threshold of visible light and in the violet-blue region. This is achieved by **doping** the mercury with, e.g., gallium (see Figure 4.9).

The lamps emit a proportion of IR radiation that may or may not be desirable - it is undesirable for heat-sensitive substrates or coatings containing highly volatile constituents. Irradiation of the object with IR can be reduced by cooling the lamps and/or reflectors with air and even water, and using IR-permeable reflectors (**selective reflectors**) [7, 43].

An interesting alternative to the normal mercury lamps (arc lamps) is that of electrodeless lamps ("fusion system"). These employ microwave radiation to excite the mercury. Advantages of these lamps include the short on and off times, their longer lifespan, greater performance per cm lamp length (up to 600 W cm^{-1}) and a higher proportion of UV radiation combined with a lower IR proportion. These advantages are offset by the technology, which, especially as regards the energy supply for the tubes, is more elaborate than in the case of arc lamps, and the length of a single lamp tube is currently limited to 250 mm. Electrodeless lamps, too, are available with different radiant spectra [44, 45].

Other types of lamp include flash lamps and excimer lamps. The former emit very strong light flashes with a broad wavelength distribution and a low IR content. Multiple lamps housed in one unit can each emit multiple flashes of 3000 W s or 3 kJ energy (typical value) per second. The extremely high intensity of the flashes leads to a good depth of cure in the films [42, 46]. Excimer lamps owe their name to the fact that an electric discharge applied to Xe and C_{12} (xenon and chlorine) produces unstable molecules, such as XeCl, which immediately disintegrate into photons. They emit virtually monochromatic radiation in the UV range, i.e. without IR fraction. This makes them attractive for curing thin layers, such as printing inks on very heat-sensitive substrates, despite their lower radiant intensity. Particularly suitable for cationic paint curing is the XeCl lamp, which emits at 308 nm [47]. Xe_2 lamps emit at 172 nm and, although they do not penetrate deeply into the film, they will excite acrylic double bonds direct, i.e. without the need for photo-initiator. This effect can be used for delustering [22, 41, 48].

The shape (curvature) of the reflectors is also crucial. Reflectors are essentially divided into **parabolic** and **elliptical** types. The former reflect the radiation in parallel, while the latter reflect into a focal plane (see Figure 4.10). Parallel rays are advantageous when the object is not completely flat and thus would not entirely pass through the focal plane of an elliptical reflector. Further reflectors with complex blend geometries are used nowadays.

Finally, it should be noted that all UV systems must be secured to prevent radiation from getting into the eyes or onto the skin.

4.3.3 UV radiant intensity and dosage

The radiant intensity emitted by the lamp is more precisely defined by the **irradiance**: $I = d\phi/dA$ ($d\phi$ radiant flux = radiant power, dA is the area perpendicularly exposed to $d\phi$), e.g. in $W\ m^{-2}$ or $mW\ cm^{-2}$. This variable is continuously measured in UV curing systems for the purposes of process control or to regulate the lamps. Great care must be taken to establish to which wavelengths this intensity applies because, in UV curing, it is very important whether maximum intensity of the radiation is at, e.g., 300 nm or 400 nm. If the (parallel) radiation impinges on an area dA of the film surface, then $S = I \cos \alpha$ is the **irradiation strength** (α angle between beam and area normal).

The degree of curing is primarily determined by the strength of the radiant energy (for a given spectral distribution) impinging on a section dA of the film surface. This radiation dose D, e.g., in $mJ\ cm^{-2}$, is given in the simplest case (object at rest, uniform irradiation) by multiplying S with the time t. Mostly, however, the object is moved at a constant velocity of 5 to more than 40 $m\ min^{-1}$ beneath one or more lamps [7]. Thus, both the incident (effective) radint flux density, I, and the radiation's angle of incidence, α, vary with time. Hence,

$$D = \int_{t_1}^{t_2} I(t) \cos \alpha(t)\, dt$$

The transport velocity $v = dx/dt$ can be used to convert the time into a distance: $dt = dx/v$, the integral of which can be expressed as

$$D = \frac{1}{v} \int_{x_1}^{x_2} I(x) \cos \alpha(x)\, dx$$

where x_1 and x_2 represents the entire transport distance over which radiation impinges on the surface dA. It can be seen that the radiation dose is inversely proportional to the transport velocity. At very high transport velocities, multiple lamps need to be arranged in a row in order for sufficient radiation doses to be achieved.

4.4 Further industrial drying and curing methods

Resistance drying and **induction drying**

This is a direct rapid heating of the metallic substrate.

High-frequency drying and **microwave drying**

This is a direct rapid heating of the wet film on a nonmetallic substrate.

Electron beam curing

This is a variant of radiation curing in which the radiation-curable film (UV paint without photo-initiator) is bombarded with electrons at 150 to 300 keV. Ultra-fast curing by free-radical polymerisation occurs; pigmentation and thickness are irrelevant. It is a very elaborate technology and thus only suitable for maximum area throughput (see Chapter 10.2.2 for more details).

Drying with dehumidified air ("low X")

Blowing layers warm dry air at 25 to 40 °C (less than 6 g of water vapour per kg dry air) over waterborne paint, e.g., prior to over-painting (with topcoat) in multi-layer systems, leads to rapid surface or pre-drying. The method is energy efficient and gentle [15].

5 General principles of film formation

The previous chapters touched on a number of technical terms that were explained wherever this was deemed necessary from the context. But the core of film formation - transformation of the applied coating material into a solid, good quality coating - was mentioned only sporadically and more or less in passing. In this chapter, we will take a much closer look at the scientific and paint-chemistry concepts behind film formation.

5.1 Polymers

5.1.1 Basic definitions

Every finished coating film contains one or more **polymers** by way of essential component. This is either identical with the binder (coalescing agent) of the coating material or is formed from the binder by chemical reaction(s) during curing. In the former case, the binder must already have a high molecular weight in the coating material prior to film formation, while, in the latter, it will consist either of much smaller molecules or oligomers, and even monomers before film formation.

Polymers (specifically: high polymers) or **macromolecular substances** consist of very large, thread-like molecules with molecular weights (M) higher than 10,000 g mol^{-1}. They are made industrially or artificially from monomers or oligomers (see below) by **polyreactions:** polyaddition, polycondensation or free-radical polymerisation (see Chapters 8.2 and 10.1). If we assume that a typical monomer has a molecular weight of approx. 100 g mol^{-1} (e.g., styrene), one polymer molecule, therefore, contains at least 100 monomer units. The number of monomer units (incorporated monomer molecules) in a polymer molecule is called the **degree of polymerisation** (P). Consequently, the size of a polymer can be expressed in terms of either M or P. Macromolecular binders can generally form sufficiently stable films by physical means, i.e. without further chemical enlargement of their molecules. Typical polymers are polymethyl methacrylate, polyvinyl chloride, polyurethane, polyester, cellulose nitrate (esterified polyglucose) and polyisoprene (natural or synthetic rubber). Polymers formed by chain-growth polymerisation, polycondensation or polyaddition are known as **chain-growth polymers**, **polycondensates** or **polyadducts** respectively.

Monomers are small molecules that generally either have two or more reactive sites in the form of **double bonds** or other **functional** (reactive) **groups** and thus can form at least linear linking (unbranched) chains or **linear polymers** by chemical end-to-end linking. The number of bonding sites offered by a monomer molecule is called its **functionality** or, less precisely, its valency. Monomers with more than two functional groups lead to **branches**, **branched** or even **crosslinked polymers**. Monofunctional monomers are sometimes used as chain stoppers in resin syntheses and as reactive diluents for conferring flexibility in paint curing. Monomers serve mainly as the raw materials for synthesising binders, but a certain proportion of monomers or traces of them are still present in many binders after synthesis. Some monomers also serve as crosslinkers or hardeners and/or as reactive diluents for paint curing. Examples are styrene, acrylates, polyamines, low-molecular-weight epoxies.

Peter Mischke: Film Formation in Modern Paint Systems

ISBN 978-3-86630-861-9

Note: The terminology here can be confusing:

- The prefix "poly" not only stands for polymers (macromolecules), but also occurs in the names of substances of different molecular size to connote "multifunctional" in the sense of two or more functionalities per molecule. Thus, polyamines, polyols, polyisocyanates and polyepoxies are substances having multiple amino, hydroxy, epoxy or isocyanate groups in the molecule.

The ambiguity is especially unfortunate in acrylic chemistry. A polyacrylate can be a polymer made by polymerisation of acrylic monomers, and a low-molecular-weight reactive diluent or oligomer with multiply linked acrylic acid (acryloyl) groups. The meaning can only be interpreted from the context.

Relatively low-molecular-weight polymers with degrees of polymerisation up to about 20 or molecular weights of at most a few thousand g mol^{-1} are also known as **oligomers**. Many binders are oligomers before film forming and are viscous or semi-solid when pure. They are transformed, possibly by curing (molecular enlargement and crosslinking), to high polymers, which are also known technically as **prepolymers**.

A great many curable coating resins have molecular weights ranging from 2,000 to 10,000 g mol^{-1}. Their physico-chemical behaviour is thus intermediate between that of the low molecular weight substances, including oligomers, and the true polymers (high polymers) and they are therefore difficult to formally describe.

Reactive diluents are liquid monomers which are added to the coating to lower or adjust its viscosity and which, unlike solvents, do not evaporate afterwards, but rather are incorporated into the film during curing. A reactive diluent is a component of the binder and is therefore not classified as part of the solvent (VOC). However, the noticeable odour of many reactive diluents clearly betrays their high volatility. Figure 5.1 explains the most important concepts again, schematically.

5.1.2 Homopolymers and copolymers

A polymer which consists of only one monomer, such as styrene, is called a **homopolymer** or, in recent terminology, a unipolymer. **Copolymers** are formed from or more different monomers.

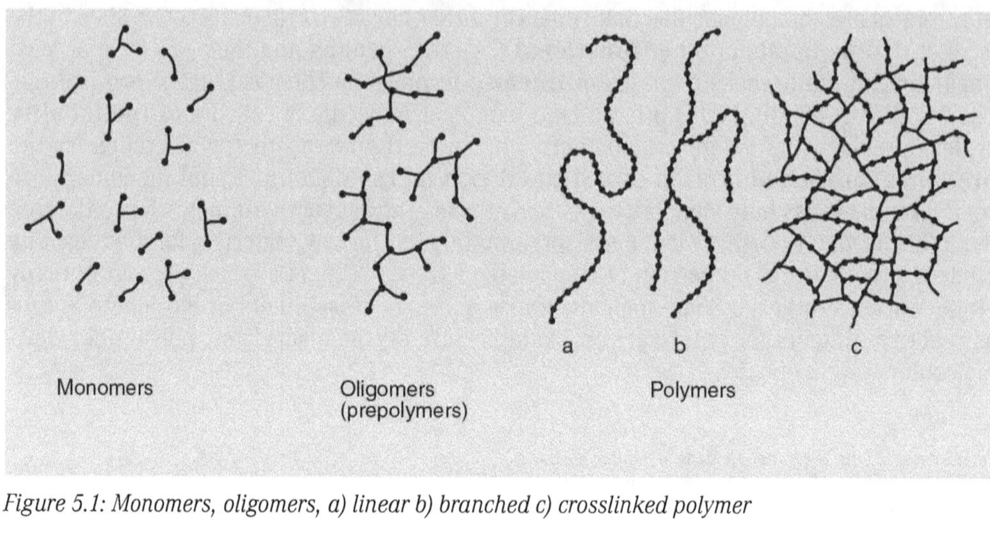

Figure 5.1: Monomers, oligomers, a) linear b) branched c) crosslinked polymer

In polycondensation and polyaddition, chain formation usually requires at least two complementary monomers, such as a dicarboxylic acid and a diol, which are incorporated alternately into the chain.The polymer in this case is also called a **homopolymer**, e.g. homopolyester, if only this minimum number of two monomers is incorporated. When three or more monomers are incorporated, the outcome is a copolymer, a **copolyester**.

Homopolymers have a very regular chain structure and thus their chains tend to pack together to the extent that they may even become **semi-crystalline**. In practice, this leads to reduced solubility and maybe to poor levelling and film haze. Paint binders are therefore almost invariably copolymers. The additional intra-molecular disorder of random copolymers (see below) completely prevents them from undergoing crystallization.

The sequence of the various monomer units in a molecule determines whether a copolymer is a

- **Random copolymer**: the monomer units are randomly distributed in the chain, i.e. without apparent regularity. Schematically:

  ...AABABBBAABABBBABABBABABBAAAABBABABBAABAAA...

- **Alternating copolymer**: two monomer units alternate regularly. Schematically:

  ...ABABABABABABA...

- **Block copolymer**: Longer segments of the same monomer units in sequence. Schematically:

  ...AAAAAAAAAAAAAABBBBBBBBBBBBBAAAAAAAAAAAAA...

- **Graft copolymer**: monomers are attached as (grafted onto) side chains to a long chain of a different monomer. Schematically:

```
...AAAAAAAAAAAAAAAAAAAAAAAAAAAAAAAAAAAAAAAA...
     B      B          B   B     B
     B      B          B   B     B
     B      B          B   B     B
     B      B          B   B     B
     B      B          B   B     B
```

5.1.3 *Average molecular weight*

In polymer chemistry, the mass and thus indirectly the size of molecules is expressed by their molecular weight. This is simply the mass of 1 mole, i.e. 6.022 · 1023 molecules. (6.022 · 1023 mol^{-1} = NA, Avogadro's constant.) An industrial polymer contains molecules of different sizes because the random nature of the polyreactions is such that the number of monomer units per molecule, i.e. the degree of polymerisation, cannot be determined to a specific value. Industrial polymers (including oligomers) therefore have a characteristic molecular weight distribution, and so only an average molecular weight can be specified for any given polymer. The most important average values for molecular weights are the number average

Equation 5.1:

$$\bar{M}_n = \frac{\sum N_i M_i}{\sum N_i}$$

N_i is the number of molecules of molecular weight M_i

and the **weight-average** molecular weight

Equation 5.2:

$$\bar{M}_w = \frac{\sum N_i M_i^2}{\sum N_i M_i}$$

The number average is the arithmetic mean of the number of molecules of different sizes. The weight average is the arithmetic mean of the weights (more accurately: the masses).

Since the weight average is given by the number of molecules multiplied by their weight, the weight average is higher than the number average. (Less experienced readers might like to verify this with a few calculations). The difference between the two is all the greater when the molecular weight distribution is wide, i.e. the more irregular the molecular size is. A measure of the **polydispersity** of the molecular weight, i.e. the width of the molecular weight distribution, is given by

$$D = \frac{\bar{M}_w}{\bar{M}_n} \qquad \text{(or just)} \qquad \frac{\bar{M}_w}{\bar{M}_n} - 1$$

Dividing the above equations by the average molecular weight of a monomer unit yields the number average ($\bar{P}_n$) or the weight average ($\bar{P}_w$) of the degree of polymerisation.

The average molecular weight and, to a lesser extent, the molecular weight distribution have a major influence on the properties of coating materials and coatings. Some of these include the paint viscosity (in the case of dissolved binders) and the two properties of film strength and flexibility. As a rough rule:

- The higher the molecular weight and the narrower the molecular weight distribution, the greater the mechanical strength of the film at a given flexibility (extensibility, elongation at break)

5.1.4 Basic types of polymer

Polymers are divided into the following groups on the basis of their physico-chemical properties:

Thermoplastics

These consist of discrete, non-crosslinked, linear or branched molecules. They soften to the point of being able to flow when heated, and are unchanged after cooling. They are soluble in suitable solvents. The binders used for physically drying paint films are thermoplastics.

Elastomers

These are loose molecular networks composed of very flexible chain fragments (network arcs). Elastomers do not melt when heated; they therefore decompose directly when exposed to elevated temperatures. They swell extensively in solvents, but are not soluble in them. Elastomeric, i.e. rubber-like, behaviour is exhibited by some sealants, for example. This state is rarely encountered – or not in the pure form at any rate – in the paints and coatings sector.

Thermosets

These are closely crosslinked, spatial molecular networks. Heating induces only a slight change in their properties. With solvents, they undergo slight, reversible swelling at best. The narrower the mesh and the more rigid the network arcs are, the harder and less flexible is the thermoset. Highly crosslinked, hard coatings contain binder in thermoset form.

5.1.5 Molecular coils

Since single chemical bonds, which are usually the most common type present in polymer molecules, can rotate about their axes and are angled with respect to the atoms, the molecules are not straight, i.e. fully extended chains, but - depending on the length, flexibility, and possibly degree of dissolution - are present rather more or less as coils (see Figure 5.2). There are extensive theories about entanglement of molecules. For practical purposes, however, the following qualitative statements usually suffice:

The **size of the coil** increases with

- increase in molecular weight
- increase in temperature
- increase in chain stiffness
- increase in solvating power of the solvent

The size of the entanglements influences the **solution viscosity** and the film strength. A further point which is crucial with regard to various polymer properties is that free polymer coils are highly diffuse, i.e. they consist largely of empty space. This allows them to penetrate each other (see Chapter 5.1.6).

5.1.6 Intermolecular forces and aggregates

Free atoms and small molecules are attracted to each other by intermolecular physical forces, which are often described as **secondary valence forces**. These can be categorized as follows [49]:

- Van der Waals forces
  - **London-Van der Waals forces** (dispersion forces) Fundamental force which is always present. Cause: quantum mechanics; oscillating dipoles. Bond energy (approx.): 8 kJ mol^{-1}.
  - **Orientation forces**:
    Electrostatic forces of attraction between permanent dipoles or between dipoles and monopoles, i.e. polar groups and perhaps ions. Weaken with increase in temperature, because thermal molecular movement counters the orientation tendency. (Ion-ion forces are ionic bonds, i.e. primary valence bonds) Bond energy (approx.): 4 kJ mol^{-1}.
  - **Induction forces**:
    Electrostatic forces of attraction between (dipoles or monopoles and the dipoles induced in neighbouring molecules by polarisation. Bond energy (approx.): 8 kJ mol^{-1}.
- **Hydrogen bonds**:
  Essentially, forces between OH or NH bonds of the O atom of one molecule and the N atom of the other molecule; schematically: Bond energy (approx.): 20 kJ mol^{-1}.

X–H········Y where X, Y = O or N
(other bonds ignored)

For comparison: A true chemical bond (primary valence) has a bond energy of the order of 300 kJ mol^{-1}.

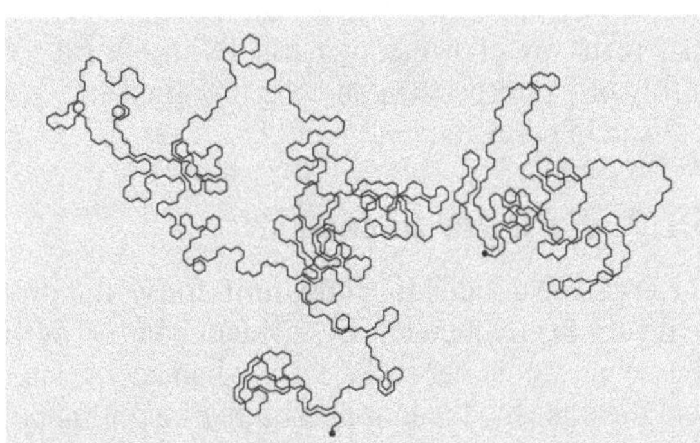

Figure 5.2: Random coil of a polymer molecule (to be considered in three dimensions); from [3]

In large molecules, such as polymers, these forces do not act over the whole molecule, but instead only locally between segments and, possibly, polar groups. However, this also leads to association and aggregation of molecules to yield what are called **aggregates**.

As explained in Chapter 5.1.5, isolated polymer molecules are present as coil. The material properties of polymers and films formed therefrom are the outcome of the physical interaction of a great many molecules. A simple calculation shows that an 1 cm^2 polymer film of 50 μm thickness contains approx. 10^{17} molecules of average molecular weight 30,000 g mol^{-1}. A rough estimate of the volume of an undisturbed coil of polystyrene, as an example, multiplied by the number of molecules shows that the total volume of the coils is about 19 times that of the film volume. This means that the coils in the film themselves are heavily penetrated and may be additionally compressed. A similar model calculation for a molecular weight of 3000 g mol^{-1} shows that the coil volume is only one to six times as great as the film volume. Coil interpenetration - defined as mutually penetrated volume - thus falls with decrease in molecular weight.

Coil interpenetration leads to a certain degree of molecular matting of the molecular threads; the scientific name for this is **entanglement**. Entanglement only occurs above a characteristic molecular weight of the polymer concerned, i.e. the molecular thread length, which is known as the entanglement molecular weight. A similar measure is the **entanglement length**, i.e. the average molecular weight between two consecutive entanglements. For many polymers, this figure is somewhere between $2 \cdot 10^3$ and several 10^4 g mol^{-1}. Oligomers or low-molecular-weight binders therefore do not form entanglements as their chains are shorter than the entanglement lengths [23]. In addition to the entanglements, coil interpenetration leads to the formation of the above-mentioned secondary valence bonds between neighbouring molecule threads, i.e. to Van der Waals bonds and, perhaps, hydrogen bonds. The desired mechanical behaviour of the film, namely tensile strength and hardness combined with high extensibility, is the cumulative effect of secondary valence bonds and entanglements. Figure 5.3 illustrates this for two isolated thread segments.

The secondary valence bonds rupture increasingly as the temperature rises; cohesion between the molecules becomes weaker, the polymer (the film) becomes softer, and finally a viscous melt forms in which the molecules slip past each other, with hardly any external force being exerted. The molten state or, at least, a greatly softened state is important in various aspects of coatings technology, such as coating powders, reflow paints (which can flow again under the of heat after they have been sanded) and thermal stripping.

Touching and interpenetration of polymer molecule coils during physical drying can cause the chain segments to associate and thus to form aggregate structures of different order - ranging from a totally unstructured "molecular felt" to the semicrystalline state. These structures are shown schematically in Figure 5.4. The cell structure in which there are hardly any interpenetrated coils and the semi-crystalline state are rarely encountered or realised in paint chemistry.

5.1.7 Polymer networks

These can be divided into **covalent**, **ionic** and **physical networks.** Covalent networks are held very firmly together by covalent bonds. They are by far the most important in coating technology. Ionic networks arise when acid groups - generally carboxylates (-COO^-) - are held together by divalent and higher-valent metal ions, such as Zn^{2+}, Zr^{2+} and Al^{3+}; this is also known as **metal salt crosslinking** . Physical networks are formed by partial crystallization, domain formation or even the aforementioned entanglements; their formation is not usually referred to as crosslinking (in the strict sense).

Networks are composed of **network arcs** and **network nodes**. The former are the linear chain segments between two network nodes, while the latter are the branching points for the chains. Network arcs, in turn, form closed lines - the network mesh. These terms are illustrated in Figure 5.5.

It is not usually possible to state the molecular weight of a crosslinked polymer (polymer network) since crosslinking yields macroscopic bodies with no distinct molecules, i.e. the molecular weights become almost infinitely large. A useful indicator critical to many film properties, however, is the crosslink density (ν). This is the number of network arcs per unit volume of the polymer. (The imprecise expression "**degree of crosslinking**" is often used as well. By this is variously meant the crosslink density, the degree of chemical conversion of the reactive crosslinking groups or the gel content (see below)). Alternatively, the average molecular weight ($\bar{M}_c$) of the network arcs can be quoted. These variables are inversely proportional to each other:

Equation 5.3:

$$\nu = \frac{n_c}{V_P} = \frac{m_C}{\bar{M}_C V_P} = \frac{m_P}{\bar{M}_C V_P} = \frac{\rho_P}{\bar{M}_C}$$

n_C is the number of moles of the network arcs
V_P is the volume of the polymer sample
m_C is the mass of all network arcs = sample mass m_P

ρ_P is the density of the polymer sample
As already mentioned in Chapter 5.1.4, an uncrosslinked polymer is thermoplastic and truly soluble in suitable solvents, whereas a crosslinked polymer cannot flow and will only swell in solvents. These property differences become particularly apparent at the **gel point**, when the paint starts to cure.

- The gel point is the point in the chemical conversion U_G of those groups

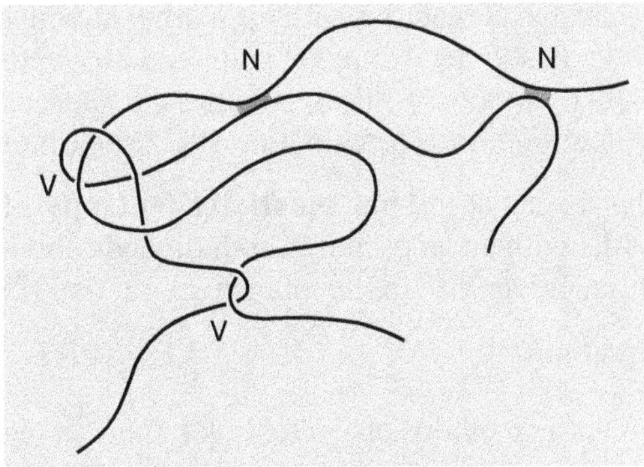

Figure 5.3: Cohesion of two threads by (S) secondary valence bonds and (E) entanglements (schematic)

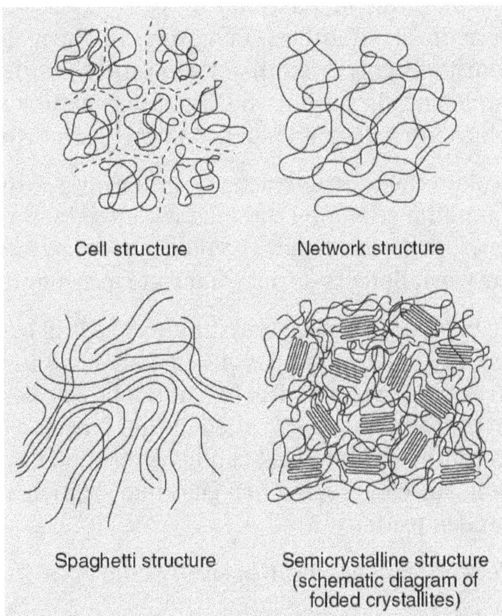

Figure 5.4: Polymer aggregates; from [3]

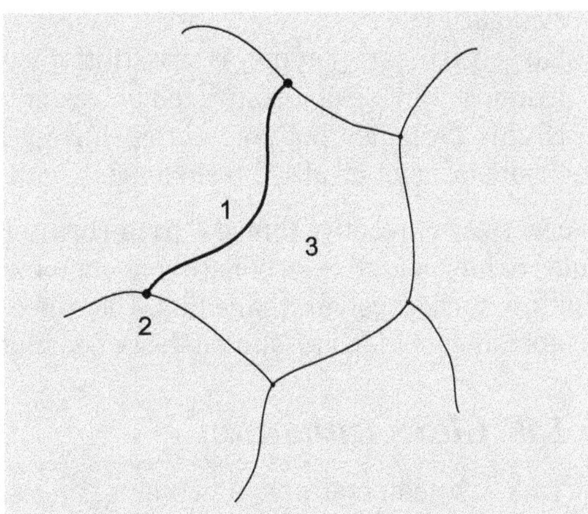

Figure 5.5: Network arc (1), network node (2) and network mesh (3) of a polymer network

capable of reacting with each other at which a network (albeit a very loose one) forms for the first time. At the gel point and above, the binder is no longer completely soluble, and further paint levelling is impossible because elastic resilience is superposed on viscous flow. (For an approximation of U_G, see Chapter 8.2.1 and Annex 3).

The theory of **rubber elasticity (entropy elasticity)** for ideal networks, which are networks without loops and free chain ends and whose node spacing follows a Gaussian distribution, gives rise to the following equation [21, 23]:

Equation 5.4: $$E = 3\nu RT$$

E is the modulus of elasticity (or Young's modulus)
ν is the crosslink density
R is the general gas constant (8.314 J mol^{-1} K^{-1})
T is the absolute temperature (Kelvin)

The **modulus of elasticity** $E = \sigma/\varepsilon$ (σ is normal stress in N mm^{-2}, ε is the strain $\Delta L/L_0$) is a measure of the stiffness or hardness of the material and increases with increase in crosslink density. Highly crosslinked paint films are hard and stiff, with low flexibility. They do not swell much in solvents. Such coatings exhibit **solvent resistance**, e.g., automotive finishes which are resistant to fuel or aircraft paint which is resistant to styrene.

Besides the crosslink density, uniformity within the film is of great importance. For maximum film strength combined with maximum flexibility, the networks must be homogeneous. Furthermore, they should not have any network arcs (loops) or free chain ends, as these **network defects** do not contribute to strength.

As mentioned earlier, crosslinking of a film begins when the gel point is reached. Of course, not all binder molecules become attached to the network by at least one bond at one fell stroke; a large mass fraction of the binder is still present by way of **sol fraction** (w_s). The crosslinked fraction is called the **gel fraction** (w_g). The following always applies: $w_s + w_g = 1$ or 100 %. The fractions can be determined by extraction (washing out) with a strong solvent, since the sol fraction goes into solution and the gel fraction is left behind as insoluble swollen matter.

Network formation can basically take place in three ways (see Figure 5.6):

- Oligomers and/or monomers react to form a great many new bonds.
  Example: 2-pack system comprising high-solids resin polyol (e.g., polyester) and polyisocyanate.
- Largely linear polymers crosslink by forming linkages between their chains.
  Example: Curing of unsaturated polyester with styrene and peroxide.
- Highly branched polymers crosslink with formation of relatively fewer new bonds.
  Example: Long-oil alkyd resin (oxidative drying).

A new class of reactive binders, **hyperbranched polymers** composed of shrub-like molecules exhibit a high level of hardness combined with unusually good flexibility after crosslinking, such as polyurethane [50]. The cause of the high performance is likely to lie in the combination of high crosslink density combined with highly mobile network arcs.

5.1.8 Glass transition

At very low temperatures, a polymer can vary from being hard and brittle to glass-like and is said to be in the **glass state**. For example, if a soft rubber article (band, tubing, etc.) is cooled in liquid nitrogen to about -190 °C, it breaks into myriad pieces when hit with a hammer. If it is steadily heated, it first undergoes weak elastification, until finally, it softens completely over a limited temperature range; at this point the polymer is said to be in the **viscoelastic state**. The precise manifestation of this state depends on factors such as crosslinked/uncrosslinked, crosslink density and molecular weight/molecular weight distribution, amorphous/semicrystalline.

Were the volume of the polymer body to be continuously determined as it was being heated up as described above and were it to be plotted against the temperature, the schematic curve shown in Figure 5.7 would be obtained. The shape of this curve stems from the following molecular factors: at a marked distance below a certain temperature called the **glass transition temperature (T_g)**, **glass temperature** or **glass point**, the polymer chains do not move at all; the only movement is vibration and rotation by atoms or groups of atoms. The coefficient of thermal expansion dV/dT, as is the case for normal solids, arises solely from the increase in these oscillations. In the vicinity of the T_g, segment oscillations are then "triggered"; these are also called microbrownian motion. The chain segments punch voids

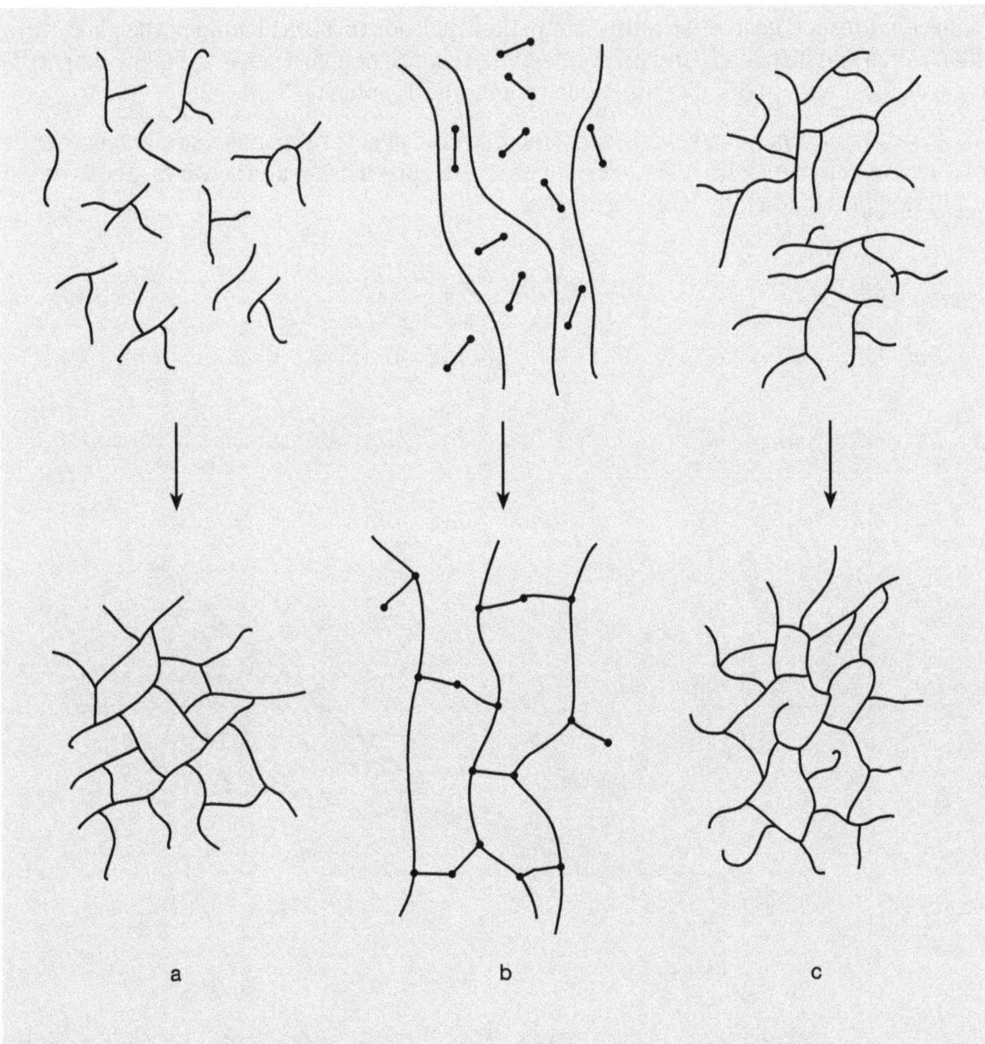

Figure 5.6: Basic ways of forming polymer networks. a) from numerous small, little-branched molecules, b) from a few large, linear molecules by crosslinking of chains, c) from large, highly branched molecules

into the polymer, the totality of which constitutes the **free volume** of the polymer. The higher the temperature, the more violent the vibrations become and the longer the oscillating segments become, because more and more of the inhibitory secondary valence bonds are being broken. The free volume increases approx. linearly with rise in temperature. If the polymer is a thermoplastic, it eventually starts to flow, as whole molecules start sliding past each other (macrobrownian motion). In the extreme case, all the network arcs in a crosslinked polymer start oscillating freely if all secondary valence bonds between the chains are thermally cleaved. The polymer exhibits rubber-elastic behaviour if the network arcs are long, and hard-elastic behaviour if they are short.

The glass transition, also known as **softening** or **freezing** depending on the direction of the temperature change, is not a phase transition like the melting of crystals or the modification of a crystal structure. It is merely a smooth transition within a temperature range of 20 to 60 K, and T_g is one temperature from this range that can only be defined accurately with an appropriate measuring method. The glass transition generally widens with increase in width of the molecular weight distribution. Where several mixed polymers of different T_g values are present, good miscibility (compatibility) leads to a broadening of the glass transition range and just one average T_g value; the existence of phase separations (incompatibilities) leads to several distinct transition ranges and T_g values [23, 51, 52].

The T_g values of copolymers, homogeneous mixtures of several polymers or homogeneous mixtures of polymers with other substances such as plasticisers and solvents are governed to a first approximation by **Fox's equation**:

Equation 5.5:

$$\frac{1}{T_g} = \frac{w_A}{T_{g,A}} + \frac{w_B}{T_{g,B}} + \frac{w_C}{T_{g,C}} + \ldots$$

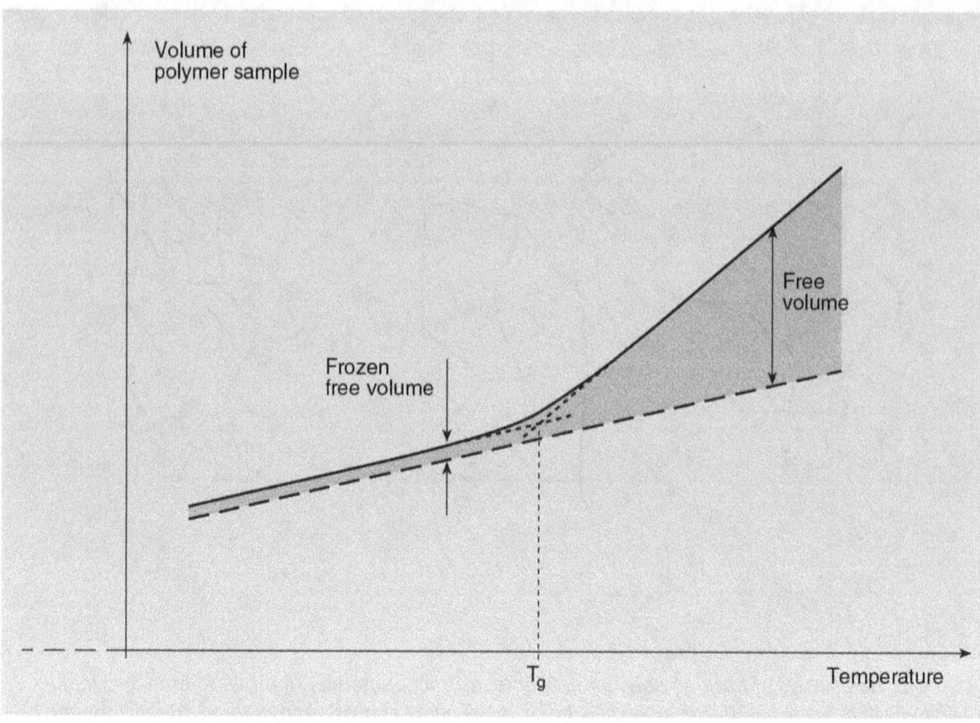

Figure 5.7: Temperature dependence of the volume and free volume of a polymer sample

$w_A, w_B, w_C, \ldots$ are the mass fractions of the substances in the mixture or monomer units in the copolymer

$T_{g,A}, T_{g,B}, T_{g,C}, \ldots$ are the T_g values of the substances in the mixture or the homopolymers of the copolymerised monomers

There are other equations, but they are complicated and for most purposes their results are not much more accurate.

Now, in addition to binders, many dried paints and other coating substances contain a preponderance of pigments and fillers. Since these are not infinitely miscible with the binder and do not have a T_g, Fox's equation naturally cannot be applied to them. Purely empirically, it should be noted that almost all pigments increase the T_g of the dry film.

The extent of the effect varies with the chemistry between the pigment particle surface and the binder and, perhaps, the additives, as well as with the particle size and distribution, the particle shape, the packing density (where there are several pigments), and the pigment volume concentration (PVC). The effect is generally 0 to 5 K per 10 % increase in PVC. The rise in T_g stems from obstruction of polymer segment mobility in the adsorption layers [23, 53].

The following five points are crucial to film formation:

- The higher the T_g of a paint film is above the current temperature (service temperature), the harder it is; conversely, the further the T_g is beneath the current temperature, the softer the paint film is.
- Fox's equation applies in principle to mixtures of polymers and solvents, too i.e. physically drying paint films, when T_g values are assigned to the solvents (solidification temperature of the supercooled, uncrystallised solvents [22]); solvents lower the T_g, i.e. they render the film softer until it can flow.
- T_g increases with increase in molecular weight. The following approximation holds:

Equation 5.6:

$$T_g = T_{g\infty} - \frac{K}{\bar{M}_n}$$

$T_{g\infty}$ is the T_g at very high molecular weight (asymptotes)
K is the empirical constant
$\bar{M}_n$ is the number average molecular weight

Since the molecular size increases during curing, the T_g and thus the hardness or the degree of dryness which it influences also rises.

- T_g increases with increase in crosslink density. Again, this is reflected directly in an increase in the hardness of the film.
- The T_g of an uncrosslinked polymer (as a concentrated solution or melt) is linked to its viscosity, η, by the **WLF equation** (Williams-Landel-Ferry):

Equation 5.7:

$$\ln\left(\frac{\eta_T}{\text{Pas}}\right) = 27.6 - \frac{40(T - T_g)}{52\ \text{K} + T - T_g}$$

T is the current temperature ($T_g \leq T \leq T_g + 100$ K)

Figure 5.8 shows a chart of the relationship for T = 293.15 K (20 °C); as this equation is so important, its derivation is provided in Annex 2.

Since the T_g increases as the film dries, WLF equation states that the viscosity also increases, a fact which manifests itself as an increase in the degree of dryness. From about 10^3 Pa·s, the state is "touch dry" and from 10^7 Pa·s, it is **"non-blocking"**[11]. 10^{12} Pa·s is roughly the viscosity of a completely dry thermoplastic at $T = T_g$ [22].

Here is a summary list showing how the T_g of a pure polymer depends qualitatively on different molecular parameters:

T_g increases with

- increase in molecular weight (up to a limit)
- increase in secondary valence forces (dipole forces, hydrogen bonds)
- increase in stiffness of the chains (e.g., by aromatic rings)
- increase in number of small rigid groups or substituents at the chains (e.g., methyl and phenyl groups)
- increase in crosslink density

Flexible sections in the main chains and long side chains lower the T_g in a process known as **internal plasticisation**.

Since numerous physical parameters of a polymer change extensively on passing through the glass region, there are also several methods for measuring the T_g. These include, above all, DMA (dynamic mechanical analysis), DSC (differential scanning calorimetry) and TMA (thermal mechanical analysis).

5.2 Solvents and polymer solutions

5.2.1 Solvents

5.2.1.1 Definition and classification

According to EN ISO 4618, a solvent is "a liquid which consists of one or more components, which is volatile under the specified drying conditions volatile and in which the binder is completely soluble." Beyond these basic technical conditions, solvents should not have any intrinsic colour, should be thermally and chemically stable (inert), have a low odour and be as toxicologically harmless as possible.

Solvent molecules are attracted to each other by the intermolecular forces described in Chapter 5.1.6. From a physico-chemical point of view, it is expedient to classify solvents according to the nature and strength of these forces:

- London-van der Waals forces only (dispersion forces):
  Aliphatic and alicyclic (cycloaliphatic) hydrocarbons (HCs) such as pentane, hexane, heptane, isoparaffins, benzines, cyclohexane, decalin.
- London and weak orientation and induction forces:
  Aromatic hydrocarbons such as toluene, xylene, solvent naphtha, "Solvesso", "Shellsol".
- London, induction and orientation forces (with separate proton donors including hydrogen bonds):
  Slightly polar molecules without cleavable protons (polar aprotic solvents), such as esters, ketones, nitro-paraffins, dimethylformamide (DMF), N-methylpyrrolidone (NMP), dimethylsulphoxide (DMSO).
- London, induction, orientation and hydrogen bonding forces:
  Polar molecules with readily cleavable protons, such as low alcohols (ethanol, propa-

[11] "Non-blocking" means that two films which have been pressed together for a defined period of time do not bond to each other

nols, butanols, amyl alcohols) and ether alcohols (butyl glycol, butyl diglycol, methoxypropanol etc).

Physically, solvents are classified primarily according to their volatility and boiling point.

5.2.1.2 *Volatility of solvents*

A substance (solvent) boils when its vapour pressure equals the ambient pressure (which is generally atmospheric pressure; about 1 bar or 10^5 Pa). Since the vapour pressure clearly depends on the temperature, every solvent has a **boiling point.** (The boiling ranges of mixed or impure solvents can be quite wide). Generally, the boiling point increases with rise in molecule size and increase in the strength of the intermolecular bonding forces, the latter being reflected in a rising **enthalpy of vaporisation**[12)]**.**

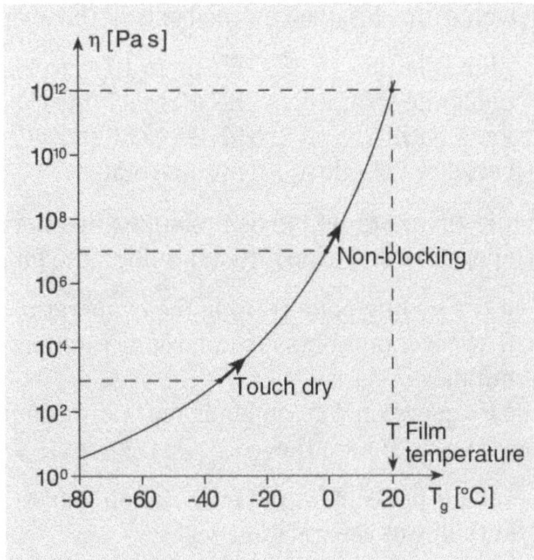

Figure 5.8: Relationship between Tg and η according to the WLF equation (for film temperature of 20 °C)

In accordance with the theory of corresponding states, the relationship between boiling point and enthalpy of vaporisation is approximated by the **Pictet-Trouton rule**:

Equation 5.8:

$$\Delta_v S_m = \frac{\Delta_v H_m}{T_s} = 84...92 \text{ Jmol}^{-1}\text{K}^{-1}$$

$\Delta_V S_m$	is the molar enthalpy of vaporisation (at T_{bp})
$\Delta_V H_m$	is the molar enthalpy of vaporisation (at T_{bp})
$T_b p$	is the boiling point (in K)

Solvents are classified by boiling point into

- **low boilers:** Boiling point/range below 100 °C
- **medium boilers**: Boiling point/range from 100 to 150 °C
- **high boilers:** Boiling point/range above 150 °C

Liquids with a boiling point above 250 °C are not solvents in the strict sense, but rather are **plasticisers** if they can penetrate and soften a polymer.

For the purposes of paint drying, more important than the boiling point is the volatility of the solvent, i.e. its rate of evaporation below the boiling point. A measure of the volatility at room temperature (23 °C) is provided by the evaporation number (EN), which indicates the time required for a solvent to evaporate under defined (standardized) conditions relative to diethyl ether (DIN 53170). High evaporation numbers indicate slow evaporation. In the USA and several other countries, the reference solvent is butyl acetate, and the inverse relationship holds: high evaporation numbers there indicate rapid evaporation.

12) The term enthalpy always stands for the heat produced or consumed in a processes (in this case, evaporation) when the process occurs under constant ambient pressure

Solvents are classified on the basis of the evaporation number as having:

- high volatility — EN up to 10 (butyl acetate system: above 3.0)
- moderate volatility — EN 10 to 35 (butyl acetate system: 0.8 to 3.0)
- low volatility — EN 35 to 50 (butyl acetate system: below 0.8)
- very low volatility — EN above 50

While there are no precise relationships between boiling point, vapour pressure at room temperature and evaporation number, the following general trends exist:

- if the boiling point is high, the evaporation number tends to be high
- if the vapour pressure at room temperature is high, the lower is the evaporation number

The evaporation rate depends on many compound-specific and external parameters. The most important are the

- vapour pressure at the evaporation temperature
- enthalpy of vaporisation
- molecular weight
- specific heat capacity
- air temperature
- air-flow rate
- contact area between air and solvent, and
- heat capacity of the carrier or substrate

One solvent is an anomaly in many respects, namely water. Its main thermal data are

- T_{bp} = 100 °C, vapour pressure (25 °C) = 24 mbar, EN = 80, specific enthalpy of vaporisation (100 °C) = 2257 kJ kg^{-1}.

A particular feature of water is its low volatility relative to its boiling point, and the very high enthalpy of vaporisation, which is many times that of organic solvents. For example, propyl acetate with T_{bp} = 102 °C has an evaporation number of 4.8 and an enthalpy of vaporisation at the boiling point of about 330 kJ kg^{-1}; it therefore evaporates about 16 times as fast as water, which has almost the same boiling point.

In practice, the low volatility of water means that waterborne paints often take longer to dry than conventional coatings. The evaporation rate is also heavily dependent on the relative humidity: the damper the air, the slower the rate of evaporation (see Chapter 6.1.1).

Standard coatings usually contain blends of solvents – even in combination with water. Raoult's law describes the overall vapour pressure p_g of a homogeneous mixture to a good approximation as follows:

Equation 5.9: $$p_v = f_1 x_1 p_{01} + f_2 x_2 p_{02} + f_3 x_3 p_{03} + ...$$

1, 2, 3, ...	are the solvents (components)
$f_1, f_2, f_3, ...$	are the activity coefficients
$x_1, x_2, x_3, ...$	are the molar fractions
$p_{01}, p_{02}, p_{03}, ...$	are the saturated vapour pressures of the pure solvents
$f_i x_i p_{0i} ...$	is the partial vapour pressure of component i

The product of f and x is the called the activity: $a_i = f_i x_i$. An **activity coefficient** is a measure of the interaction of a component with the rest of the system. Values less than one indicate that the intermolecular interaction (bond strength) is stronger than in the pure state; the partial vapour pressure is then depressed. For values greater than one, the opposite is true.

Unfortunately, activity coefficients are not known with precision nor are they predictable. It should also be recalled that the evaporation rate is not exactly proportional to the vapour pressure.

Another constraint on the predictability of evaporation processes is that the solvents evaporate from the mixture at different rates due to their different volatilities, as a result of which the composition starts to change from what it was initially. Consider the example of a 1 : 1 mixture of butyl glycol/water at its boiling point. First, butyl glycol has a higher vapour pressure and therefore evaporates faster than water. Thus, the concentration of water increases, which, according to Equation 5.8, leads to higher water vapour pressure and thus to a higher rate of water evaporation. Finally, a composition is reached at which both components evaporate constantly at the same rate and the composition of the mixture no longer changes. This mixture is called an **azeotrope** (or azeotropic mixture), and in this case has the following composition: butyl glycol/water = 20.8 : 79.2 (mass fractions) and boils at 170 °C, i.e. at a temperature lower than butyl glycol ($T_{bp} \approx 170$ °C) and water itself [54]. Such azeotropes also form when evaporation occurs below the boiling point; in that event, the composition is different from that at the boiling point and, if water is also present, also depends on the relative humidity.

And it gets even more complicated: the evaporation rates and azeotrope compositions are also dependent on the interaction with the binder. Solvent components strongly attached to the binder evaporate more slowly than weakly attached ones (see Chapter 5.2.2.3).

5.2.2 Polymer solutions

5.2.2.1 General

If a sample of an uncrosslinked, amorphous polymer is introduced into a suitable **solvent**, the solvent molecules penetrate into the coils, enveloping the polymer molecules to form secondary valence bonds. The latter process is called **solvation.** To an extent depending on the molecular weight of the polymer molecules, the solvating power of the solvent for the polymer present and the volume ratio of polymer/solvent, dissolution more or less leads to widening of coils and their disentanglement or separation. This process continues until a homogeneous system, i.e. one of uniform composition, or true polymer solution has formed. Dissolution here - in contrast to dissolution of crystalline substances, such as that of common salt in water - is not accompanied by crystal degradation, and thus should be understood as merely being a mixing process. For example, when the dissolution process is reversed, i.e. the solution is boiled down, a separate phase does not usually **precipitate** out of solution (see below for an exception), but rather the system gradually solidifies. This is precisely the process which occurs when conventional paints dry physically.

The solvent located in the coils can be classified (very schematically) as bound and free solvent. **Bound solvent** adheres to the molecular threads so strongly that it moves with them and is not rinsed off. **Free solvent** is virtually unbound and washes through the coils as they undergo translation through the solution (see Figure 5.9). When a film undergoes physical drying, the free solvent is released much more readily than the bound solvent.

If increasing amounts of a non-dissolving solvent or non-solvent are now admixed to this solution, it will cause turbidity once it reaches a certain concentration. The polymer starts to precipitate out of solution in swollen form, i.e. starts to become incompatible with the rest of the solution. The molecules of highest molecular weight precipitate out first because the solubility of polymers generally decreases with increase in molecular weight; upon further addition of solvent, they are followed by the next smallest molecules, and so on.

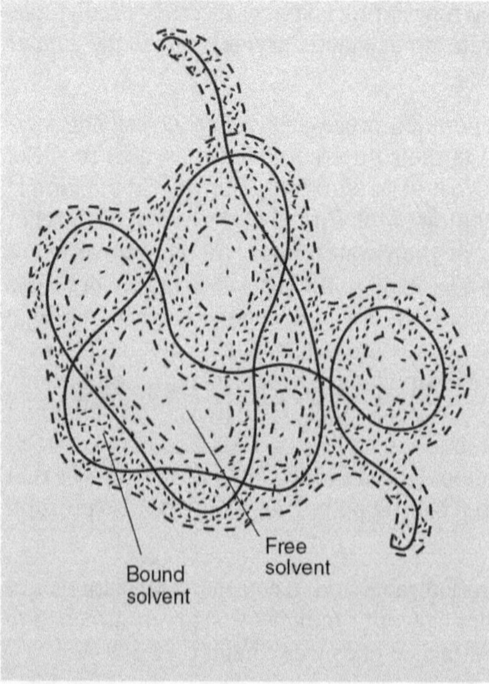

Figure 5.9: Bound and free solvent in a partially washed-through polymer coil (highly schematic)

This process is of huge importance to paint drying, because if the beneficial solvent evaporates too quickly from the solvent blend in the wet film, the binder may precipitate and so defects, such as levelling problems, may occur.

For the sake of completeness, reference is made to the term theta state, which is still encountered occasionally. This state of any given polymer in a dilute solution is determined by the combination of polymer type, solvent and temperature, at which all the solvent is bound and thus the coils are not washed-through, and at which the amount of polymer is just at the solubility limit. A theta solvent is thus a relatively poor solvent for the polymer present.

Crosslinked polymers can basically only swell, but cannot form solutions, as all molecular threads are bound together chemically, i.e. are not free to float around.

Solutions must not be confused with **dispersions** and **emulsions**, which are disperse, micro-heterogeneous systems consisting of polymer particles or oligomer droplets ranging from 0.03 to several µm in size and which are uniformly distributed in a non-solvent, usually water, acting as dispersant (see Chapter 5.3).

5.2.2.2 Affinity between polymer and solvent; and solubility parameters

The ability of a solvent to dissolve or swell a given polymer depends mainly on the strength of the interaction between the polymer chains or their monomers and the solvent molecules. The Flory-Huggins solution theory is a simple but semi-quantitative and accurate model that describes this polymer interaction in terms of χ, the Flory-Huggins interaction parameter. This can have values between 0.5 and zero, with 0.5 denoting weak polymer-solvent interaction (equivalent to the theta state) and zero standing for strong interaction. The Flory-Huggins theory can be used to derive a more practical variable that is commonly employed in paint chemistry – the solubility parameter.

The Hildebrand solubility parameter, δ, is the square root of the **cohesive energy density** of the substance in question, i.e. the polymer or solvent in this case:

Equation 5.10:

$$\delta = \sqrt{\frac{E_k}{V}}$$

E_k is the cohesive energy,
V is the volume of the sample

The cohesive energy density is the amount of energy which must be applied to this substance to completely separate all the molecules. This separation process is identical with evaporation of the sample and so δ for evaporable substances can be calculated from the energy of vaporization. For non-evaporable materials, such as polymers, an indirect approach is needed (see below). The unit employed for δ nowadays is $(J\ cm^{-3})^{1/2}$, it is 2.05 times as large as the previous unit $(cal\ cm^{-3})^{1/2}$.

The **enthalpy of solution**, $\Delta_{mix}H$, i.e. the enthalpy change associated with the dissolution of a polymer, computes to

Equation 5.11:

$$\Delta_{mix} H = V_S\, \Phi_P\, (\delta_P - \delta_S)^2$$

V_S is the volume of the solvent (S)
Φ_P is the volume fraction of the polymer (P)
δ is the solubility parameter

It is clear that the enthalpy of solution in this model can only be greater than or equal to zero. A positive value, however, signifies an endothermic reaction, which is thermodynamically unfavourable. The reverse conclusion then leads to the following rule:

- The closer δ_P and δ_S lie together, i.e. the lower the endothermic nature of the dissolution process, the greater is the miscibility of two substances or the solubility of a polymer in a solvent. (The more favourable case of an exothermic reaction is mathematically impossible here.)

The aforementioned Hildebrand **unidimensional solubility parameters** often lead to wrong predictions about solubility. Hansen **three-dimensional solubility parameters** are used almost exclusively nowadays instead. These are obtained by splitting the cohesive energy into a dispersive (from London-van der Waals forces), a polar (from the orientation and induction forces) and a hydrogen bond contribution. This yields the following definition:

Equation 5.12:

$$\vec{\delta} = \vec{\delta}_D + \vec{\delta}_P + \vec{\delta}_H$$

and

$$\delta = \sqrt{\delta_D^{\,2} + \delta_P^{\,2} + \delta_H^{\,2}}$$

δ is the overall parameter
δ_D is the disperse parameter (contribution)
δ_P is the polar parameter (contribution)
δ_H is the hydrogen bond parameter (contribution)

In charts, δ_D, δ_P and δ_H are plotted at right angles to each other. The vector δ then points from the origin to the point (δ_D, δ_P, δ_H). In other words, the three-dimensional solubility parameter of a substance can be plotted as one point in the δ_D-δ_P-δ_H coordinate system. The difference between the solubility parameter of the polymer and solvent in space, which, as stated above, is a measure of the (thermodynamic) solvating power of a solvent for a given polymer or binder, may be computed from Pythagoras' spatial theory or the rules of vector algebra as follows:

Equation 5.13:

$$\Delta\delta = \sqrt{(\delta_{D,P} - \delta_{D,S})^2 + (\delta_{P,P} - \delta_{P,S})^2 + (\delta_{H,P} - \delta_{H,S})^2}$$

Δδ is the difference in solubility parameters
$\delta_{D,P}$ $\delta_{D,S}$ are the disperse parameters of the polymer or solvent
$\delta_{P,P}$ $\delta_{P,S}$ are the polar parameters of the polymer or solvent
$\delta_{H,P}$ $\delta_{H,S}$ are the hydrogen bond parameters of the polymer or solvent

Physical drying depends critically on the direction in which the solubility parameter of the current solvent mixture shifts, relative to the parameter of the binder, when the solvent components exhibit dissimilar evaporation. The following approximation holds for the parameters of solvent mixtures:

Equation 5.14: $\delta_D = \Phi_1\,\delta_{D,1} + \Phi_2\,\delta_{D,2} + \Phi_3\,\delta_{D,3} + \ldots$ (similarly for δ_P and δ_H)

$\Phi_1, \Phi_2, \Phi_3, \ldots$ are the volume fractions of the solvents 1, 2, 3, ...

The estimation, however, becomes less accurate as the number of different solvents present increases.

The solubility parameter of a polymer or binder whose cohesive energy cannot be measured through evaporation, can be determined experimentally as follows:

Dissolution trials are conducted on a large number of different solvents or solvent mixtures of known parameters. A plot of solvent parameters is made as cloud points on the three-dimensional system. The δ_D parameter is perpendicular to the paper plane. Since the δ_D values do not differ greatly from each other, the mean of close values is usually taken as the parameter, as a result of which two-dimensional sections are created which are parallel to the plane of the paper and pass through the solubility parameter space. A line is drawn in the system along the boundary between solvents and non-solvents. All solvents now lie within the **solubility region** and all non-solvents lie outside it. The two-dimensional plot returns planar solubility regions for the various δ_D values. The following now holds:

- the three-dimensional solubility parameter of the polymer (binder) is located in the centre of gravity of the solubility region.

Thus, the parameters (or parameter points) of thermodynamically good solvents are close to the centre, while those of poor solvents are at the boundary and those of non-solvents are outside the solubility region. Empirically, it has been found that 6 $(J\ cm^{-3})^{1/2}$ is the difference in spatial parameters above which solubility usually no longer occurs. Where the parameters for binding agents and solvents are known, this rule of thumb can be used to make **solubility predictions**.

Figure 5.10 shows the **solubility parameter diagram** for a melamine resin, and Table 5.1 lists some examples of solubility parameters.

5.2.2.3 Vapour pressure of a solvent in a polymer solution

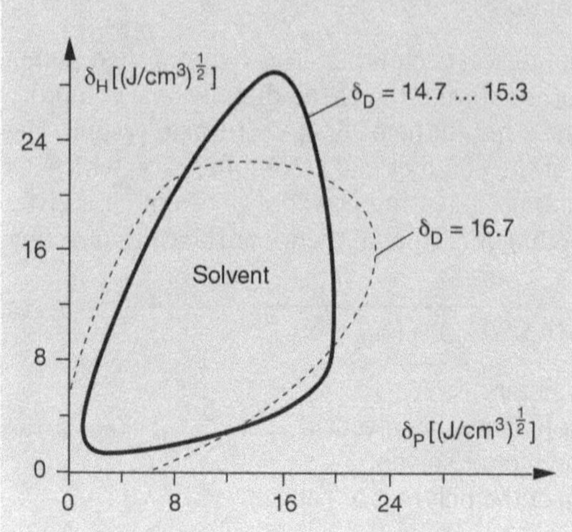

Figure 5.10: Solubility parameter diagram of a melamine resin; from [3]

Raoult's law, namely $p = p_0\,a = p_0\,f\,x$ (where p_0 is the vapour pressure of the pure solvent), states that a solvent in a homogeneous mixture with other low-molecular-weight substances, has a vapour pressure p, which is proportional to the mole fraction x and the activity coefficient f of the solvent in the mixture. (For the pure solvent, $x = f = a = 1$). This is true – in a somewhat modified form – for polymer solutions, too. In these, the solvent is diluted to an extent by the polymer, a fact which formally leads to a smaller x, and thus to a lower vapour pressure. In addition, there is the solvent-polymer interaction, which is reflected in f. The stronger it is, the lower is f and thus p. In other words:

Table 5.1: Solubility parameters of selected solvents and binders in (J cm⁻³)$^{1/2}$
[1] from [3]
[2] from [6]
[3] from [13]
[4] unidimensional parameter, calculated from the contributions in Equation 5.12

Substance	δ_D	δ_P	δ_H	δ[4]
n-Hexane[1]	14.9	0	0	14.9
Toluene[1]	18.0	1.4	2.0	18.2
Methyl isobutyl ketone[1]	15.3	6.1	4.1	17.0
Butyl acetate[1]	15.8	3.7	6.3	17.4
Isobutanol[1]	15.3	5.7	15.8	22.7
Water[2]	14.3	16.3	42.6	47.8
Hydrocarbon resin[1]	17.6	1.2	3.6	18.0
Long-oil alkyd resin[1]	20.4	3.4	4.6	21.2
Polymethyl methacrylate[2]	18.7	10.1	8.5	22.9
Cellulose nitrate[3]	15.4	14.7	8.8	23.0
Hexamethoxymethyl melamine (HMMM)[1]	20.4	8.5	10.6	24.5
Epoxy resin[2]	17.3	11.2	11.5	23.5

- good solvents tend to be retained in the paint film more than poor ones as it dries.

However, this effect may possibly be overcompensated by the higher volatility of the good solvent, In that event, the poor solvent accumulates and film formation is impaired due to precipitation of the binder.

5.2.2.4 *Viscosity of polymer solutions*

Any discussion of the viscosity of polymer solutions or binders must distinguish between dilute and concentrated solutions. Dilute solutions contain solvated molecules or macromolecular molecule coils which are spaced apart and do not touch or penetrate each other. Relatively simple laws govern this state, e.g. the **Staudinger-Mark-Houwink equation (SMH equation)**. However, paint technology and thus also film formation, deals with concentrated solutions, wthere "concentrated" means that the coils touch each other and are more or less hooked together or even penetrating each other. For high-molecular-weight polymers, this state can be achieved at a concentration of less than 1 g l⁻¹ [13].

The important aspects here as regards film formation are that the viscosity of concentrated polymer solutions depends on the molecular weight, concentration and temperature. The relationship between viscosity, molecular weight and concentration can be described by several empirical equations, each of which is of limited scope. One equation frequently encountered in the literature is [3, 7, 22, 26].

Equation 5.15:

$$\ln\frac{\eta}{\eta_0} = k\,w\,\sqrt{M}$$

η is the viscosity of the solution
η_0 is the viscosity of the pure solvent
w is the mass fraction of the polymer (corresponds to nonvolatile fraction)
M is the molecular weight (mean)
k is a constant

There is no consistency in the literature concerning the type of average molecular weight to employ. The equation is said to apply best to oligomers of narrow-molecular-weight distribution in good solvents over the range 0.01 to 10 Pa s [22]. For higher-molecular-weight binders, the exponent is greater than 0.5.

This equation shows first that, to achieve lower solution viscosities or high concentrations of binders, such as in high-solids coatings, the solvent should have the lowest-possible viscosity. In concentrated polymer solutions, the polymer coils progressively weaken their grip on each other as the secondary valence bond sites become more saturated by solvent. Moreover, the coils should be as small as possible, i.e. not very extended, a fact which is not the case for good solvents with δ parameters similar to those of the binder. For alkyd/melamine resin combinations, e.g., it has been shown in agreement with these deliberations that low paint viscosities result from solvents whose solubility range in the solubility parameter diagram is located at the boundary defined by high δ_P and δ_H values [55]. The aforementioned resins are relatively polar and form intermolecular hydrogen bonds via OH groups. Polar solvents, i.e. hydrogen bond acceptors, such as esters and ketones, or hydrogen bond acceptor-donors, such as alcohols, can rupture these hydrogen bonds by forming their own hydrogen bonds with the binder molecules and separating the molecules in the process. The outcome is a drop in viscosity.

The viscosity of a polymer solution or melt decreases with increase in temperature. For polymer solutions at temperatures above T_g +100 K (of the solution) and low molecular weight, nonassociated liquids, the Arrhenius equation (Equation 2.1) is a fairly good approximation. For concentrated polymer solutions and polymer melts in the temperature range T_g to T_g + 100 K, the WLF equation (Equation 5.7) is usually more accurate [22, 56, 57].

5.3 Aqueous-disperse systems

5.3.1 Basic terms and classification

By the **disperse state** of a substance is meant its microscopic or sub-microscopic distribution form in a solvent or dispersing agent. In the following discussion, the solvent or dispersing agent is always taken to be water.

In principle, a distinction must be drawn between solutions and dispersions:

- In a **solution** , the binder molecules are distinct (molecularly disperse) or slightly associated. Solutions are clear to very slightly turbid (opalescent).
- A **dispersion** contains colloidal[13] or coarser particles or droplets composed of many binder molecules which are distributed in the dispersing agent (here: water). In contrast to solutions, dispersions have essentially sharp phase boundaries between the dispersing agent and the substance dispersed in it (the disperse substance). Where the dispersed particles are droplets, the dispersion is called an **emulsion**. Dispersions are generally more or less turbid, the turbidity generally increasing with increase in particle or droplet size and increase in the difference between the refractive index of water and the disperse phase.

[13] The colloidal particle-size range extends from 5 to 50 nm

There are also **hybrid systems**, which in this context are mixtures of binder solutions and dispersions, generally primary dispersions.

Dispersions, in turn, are divided into primary and secondary types:

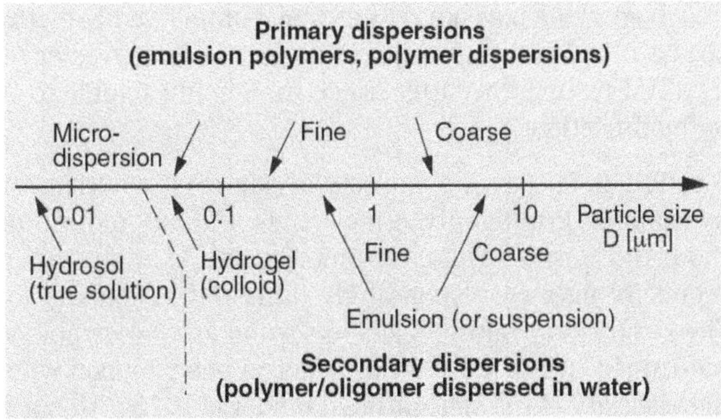

Figure 5.11: Disperse states of binding agents in water, as defined in industry; adapted from [3, 4]

- A **primary dispersion** is formed by direct emulsion polymerisation of simple monomers to yield a distribution in water of colloidal polymer particles that have a high molecular weight. Other names for this are **polymer dispersion**, **emulsion polymer** and **latex**.
- A **secondary dispersion** is created by dispersing or emulsifying a polymer or oligomer in water. Examples: polyurethane dispersion, alkyd resin emulsion.

Figure 5.11 gives a schematic overview of aqueous-disperse binder systems.

5.3.2 True and colloidal solutions

For polymer molecules to be truly soluble in water, they must be very hydrophilic (water-loving). With a few exceptions, conventional, unmodified paint binders are hydrophobic and insoluble, and therefore need to be chemically modified. This modification consists in incorporating a sufficient quantity of acidic or basic groups, carboxy (COOH-) or amino (NR_2, NHR, NH_2) groups into the binder molecules to yield highly hydrophilic centres upon **neutralisation** with low molecular weight bases or acids (see Figure 5.12). The neutralisation of carboxy groups with amines or ammonia requires a pH between 7.5 and 8.5 if a high **degree of neutralisation** is to be obtained. Too little neutralisation harbours the risk of an uncontrollable increase in viscosity or even binder precipitation, e.g. if atmospheric CO_2 continually acts on the solution or the paint, some of the amine partially evaporates prematurely or enters into a chemical reaction. Amine binders are neutralised for cathodic electrodeposition with acetic acid, and sometimes also with formic, propanoic or lactic acid. The pH value must be between 5 and 6.5 [3, 58].

Additionally or by way of partial alternative, the molecules can also contain non-ionic hydrophilic elements, such as polyethylene oxide chains (...-O-(CH_2-CH_2-O)$_n$-CH_2-CH_2-O-...) comprising 5 to 50 ethylene oxide units. Furthermore, on account of the manufacturing process and to increase the solvating power, the solutions may also contain substantial amounts of fully or partially water-miscible solvents

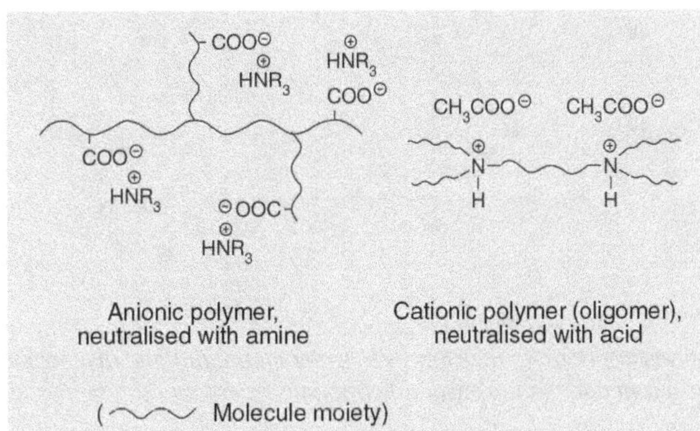

Figure 5.12: Examples of ionically dissolved binders; from [3]

(cosolvents,, such as butyl (di)glycol, butanol and N-methylpyrrolidone. The solutions could be, e.g. alkyd resins, acrylic resins, saturated polyester resins and special melamine resins (HMMM resins); the latter range from being soluble to self-emulsifying, without the need for modification

A condition for true, i.e. molecularly disperse solutions of binders, is the presence of many hydrophilic groups in the molecules and low molecular weights of no more than a few thousand g mol^{-1}. Thus, for anionic stabilisation, the **acid number**[14)] must exceed 40, representing a correspondingly high amine demand for the purposes of neutralisation. The strong ionic nature leads in turn to an undesirable viscosity anomaly called a **"water mountain"** (see Figure 5.13). This is a phenomenon whereby, following addition of amine, the viscosity of the concentrated binder (e.g. 70 % in butyl glycol) supplied in, e.g., a water-miscible organic solvent or solvent mixture, exhibits a more or less pronounced rise in solution viscosity upon dilution with water. The height and position of the water mountain depend on the chemical nature of the binder, its molecular weight, its acid number and the type and quantity of amine used for neutralisation. Although its height can be reduced by further adding the best possible cosolvent, such as butyl glycol, this of course leads to an increase in the VOC value and is therefore counter-productive. Theories about the origins of the water mountain are essentially consistent, but these will not be discussed here [instead, see 4, 22, 59]. With regard to film formation, the water mountain – if present at all – is important because the paint or binder solution becomes more concentrated during drying and so also "passes through" the water mountain. As outlined in Chapter 3, the resulting viscosity anomalies can greatly influence levelling and sagging.

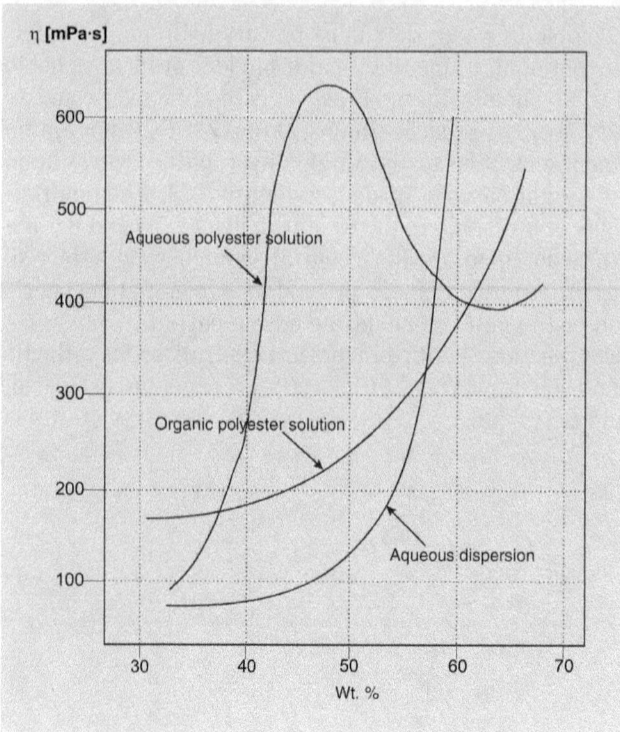

Figure 5.13: Viscosity of polyester dissolved in organic solvent, polyester dissolved in water (with water mountain) and dispersion as a function of the nonvolatile fraction; from [60]

Owing to the

- high amine demand
- high content of cosolvent or the water mountain
- the low molecular weight (weak physical drying only) and perhaps
- vrapid hydrolysis (saponification) of polyester-based binders in aqueous solutions or paints in storage

true binder solutions are now only of secondary importance [15].

With falling content of acid groups and thus of amine, and increasing molecular weight (up to approx. 10^5 g mol^{-1}), the true solubility of the binders transitions into colloidal dispersibility or self-emulsifiability. It can be supported by internal or external **emulsifiers**, such as polyether

14) The acid number is the mass of potassium hydroxide (KOH) in mg that is required for neutralisation of the acid (groups) contained in 1 g of substance

chains and surfactants. Where stable, colloidal solutions of coils of associated binder molecules are present, the solutions are almost clear or opalescent. These systems, also known as **hydrogels**, are the direct modern descendant of the true soluble binders[15] [4, 15]. The advantages over real molecularly-disperse systems are

- lower cosolvent and amine demand
- no water mountain
- better hydrolytic stability
- higher molecular weight (better initial physical drying)

5.3.3 Primary dispersions

Primary dispersions, such as pure acrylate, styrene acrylate, vinyl acetate copolymer and styrene-butadiene dispersion are obtained direct by **emulsion polymerisation**. They consist of a water phase which contains **emulsifiers** (surfactants) and perhaps **protective colloids**, such as polyvinyl alcohol or cellulose derivatives, and further auxiliaries, into which micro-structured polymer particles of nearly spherical shape are dispersed.

Primary dispersions can be broadly characterised as follows:

- average molecular weight (M_n): 10^5 to 10^6 g mol^{-1} (in some cases, higher)
- mean particle diameter: 0.03 to several μm
- nonvolatile fraction: 30 to 60 % (in some cases, higher)
- viscosity: from almost as thin as water to syrupy (protective colloids have a thickening action)
- with increase in particle size: translucent -> shimmering bluish-white, almost opaque -> completely white, opaque
- thermodynamically metastable, i.e. of limited shelf life and of limited sensitivity to chemical, mechanical and thermal effects.

A major advantage of polymer dispersions over polymer solutions is that their viscosity is low, despite very high molecular weights (over 100,000 g mol^{-1}) at medium solids levels. The concentration dependence of the viscosity of dispersions (including, pigment dispersions) is described by the **Mooney equation** [18]:

Equation: 5.16:

$$\ln\frac{\eta}{\eta_k}=\frac{K_S\,\Phi_D}{1-\frac{\Phi_D}{K_P}}$$

η is the viscosity of the dispersion
η_K is the viscosity of the dispersing agent (water phase)
Φ_D is the volume fraction of the disperse phase
K_S is the shape-dependent factor (2.5 for spheres)
K_P is the packing factor (0.637 for densely packed spheres)

For the purposes of film formation, the only interesting part of this equation is the denominator. If the quotient Φ_D/K_P approaches the value of 1 due to increasing concentration (drying), there is a steep rise in the viscosity, η (see Figure 5.13). The particle spheres come into touch and strong interactions immediately occur.

[15] The term hydrogel is not used consistently in the literature. This applies even more so to the term hydrosol, which is why this term is not used in this book

Primary dispersions largely undergo purely physical film formation by deformation and **coalescence** of the particles. The films are really stable films because of the high molecular weight. Coating materials based on primary dispersions are called **emulsion paints** and **dispersion varnishes.** The former have PVC values ranging from 40 to 90 %; supercritical indoor paints are porous and dull, while the subcritical dispersion varnishes are paint-like. Dispersion varnishes usually have PVC values below 25 % and, as clearcoats, can be totally free of pigment.

5.3.4 Secondary dispersions and emulsions

Secondary dispersions are micro-heterogeneous, i.e. more or less turbid dispersions of polymers in (mostly) water, which arise through dispersing or even simple stirring of initially anhydrous binders. Organic auxiliaries added during synthesis or prior to dispersion for the purpose of lowering the viscosity of the resin can be removed – if desired and feasible – by distillation to yield completely solvent-free dispersions. If the disperse substance, i.e. the binder, is a liquid, the result is an emulsion. In a broader sense, emulsions are counted as dispersions, especially since many binders are semi-solid in the disperse state, i.e. are neither pure liquid nor pure solid.

Secondary dispersions have much lower molecular weights than primary dispersions because the polymer must be dispersible. In a great many properties, secondary dispersions rank between colloidal solutions (hydrogels) and coarse primary dispersions, it not being possible or meaningful to make a sharp distinction from colloidal solutions. The hydrophilic nature of the molecules determines whether the products are thermodynamically stable, amine-neutralised, almost clear products, or essentially milky, emulsifier-containing, coarser systems of limited storage stability. The latter often contain other auxiliary components, such as protective colloids and organic solvents.

5.4 Diffusion

By diffusion is meant the migration of a dissolved substance within a solution due to a concentration difference or the spontaneous mixing of one substance with an adjacent one. Examples include the gradual spread of ink drops in still water and the swelling of a paint film by drops of solvent.

The diffusion rate in physical chemistry is given by the flow of substance across a unit area; it is traditionally known as the mole flux density and is proportional to the concentration gradient or the negative concentration slope -(dc/dx) of the diffusing substance. The **diffusion coefficient**, D, serves as the proportionality factor. The corresponding equation is called **Fick's first law**.

Equation 5.17:

$$j=\frac{\dot{n}}{A}=-D\frac{dc}{dx}\approx-D\frac{\Delta c}{\Delta x}$$

j is the mole flux density across a unit area
$\dot{n}$ is the mole flux dn/dt
A is the area plane across which diffusion takes place
Δc is the concentration difference (in mol L^{-1})
Δx is the distance over which the concentration difference exists
D is the diffusion coefficient

The rounding symbol indicates that the integral gap does not match the differential (local) gap exactly, as the concentration profile c(x) is usually not strictly linear. Moreover, strictly speaking, the activity difference should be used instead of the concentration difference. For practical considerations and assessments, though, the simplification above is adequate.

For a given concentration gradient, the diffusion rate is expressed by the diffusion coefficient. For dilute solutions

Equation 5.18:

$$D = \frac{RT}{N_A \, f_R}$$

R is the general gas constant (8.314 Nm mol^{-1} K^{-1})
T is the temperature T (absolute temperature) in Kelvin
N_A is Avogadro's constant ($6.022 \cdot 10^{23}$ mol^{-1})
f_R is the frictional resistance of a molecule of the diffusing substance

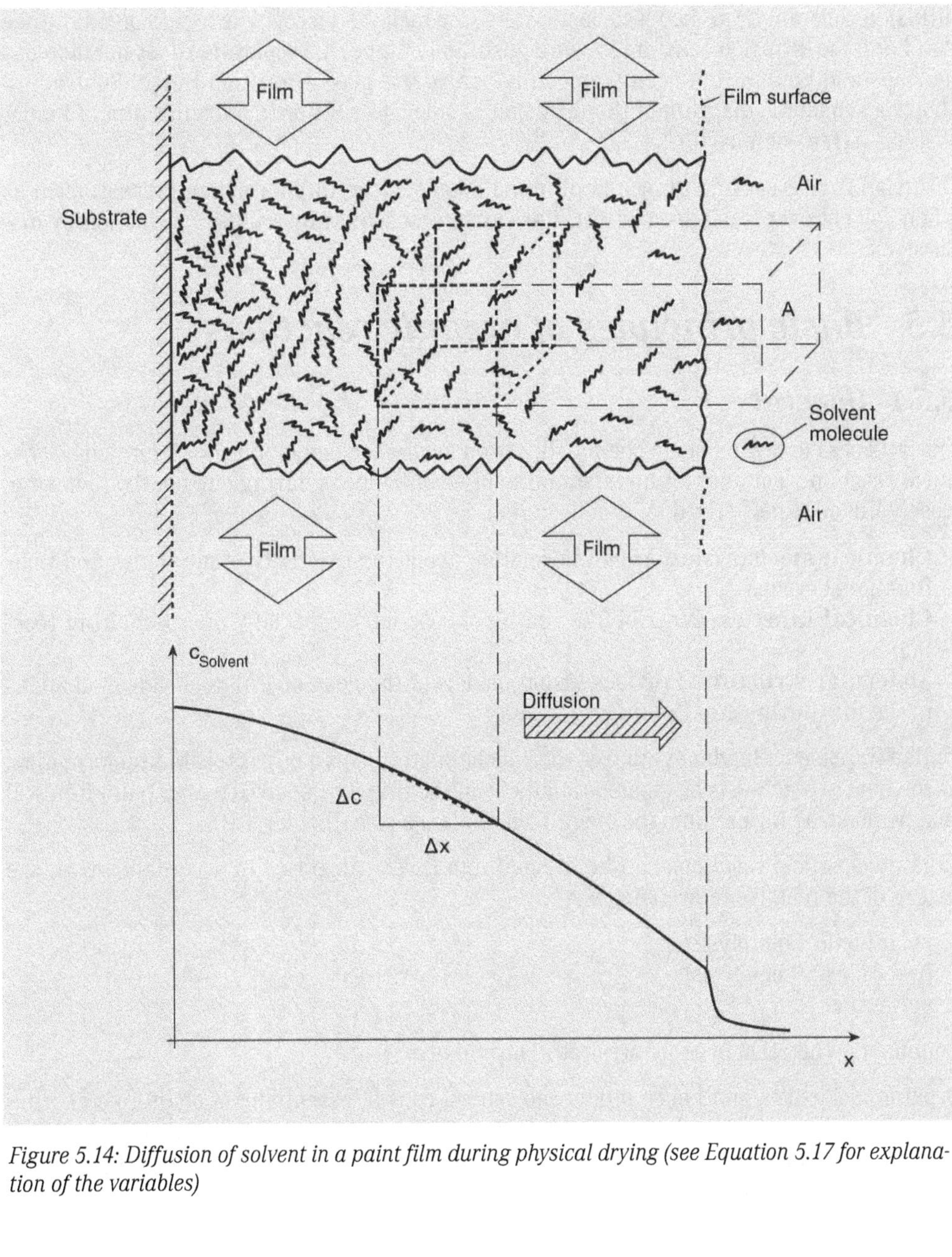

Figure 5.14: Diffusion of solvent in a paint film during physical drying (see Equation 5.17 for explanation of the variables)

The frictional resistance f_R is the ratio F/v of the frictional force and migration rate and is proportional to the viscosity and the effective (equivalent) molecular radius: $f_R = 6\pi\eta\, r_{eff}$. In other words, large molecules diffuse more slowly than small ones. And, since in Equation 5.18, the numerator increases with T while the denominator η (and with it f_R) decreases, the diffusion rate increases markedly with increase in temperature.

With regard to the rate of diffusion of, e.g., solvents or the inter-diffusion of polymer molecules in a thermoplastic or at best weakly crosslinked polymer (film), the key role is played by the free volume. The many submicroscopic voids in the polymer compound render these more permeable to molecules or the ends of molecular chains. It has been found empirically that the diffusion rate increases with the free volume and this, in turn, increases with increase in the difference $T - T_g$, i.e. with T at constant T_g (see Chapter 5.1.8 and Annex 2). T_g represents a major border. It separates the glass state of very high viscosity and very small diffusion constant from the viscoelastic state in which the viscosity is orders of magnitude lower and the diffusion constant is commensurately high. This separation does not increase by leaps and bounds, but occurs smoothly across the glass transition range. Figure 5.14 shows a schematic diagram of the diffusion of solvent or water in a drying film of binder that has a free volume ($T > T_g$).

Diffusion is of great importance at different stages of the film formation process, such as when solvent evaporates from a wet film, during crosslinking, and film formation by dispersions.

5.5 Basic principles of organic reactions

5.5.1 General

The bulk of a coating exists in chemically cured form, i.e. a polymer network formed by chemical reactions. For a better understanding of what happens during curing, the following, closely linked aspects need to be considered:

- **Chemical-mechanisms**: Yield information about the reactivity of molecules and their functional groups.
- **Chemical kinetics**: Describe the timing of reactions, especially of crosslinking reactions.
- **Molecular structure**: Provides visual models of the geometrical (topological) changes in the film during crosslinking.

Unlike the case for binder synthesis, thermochemistry plays a minor role in film formation, as the heat of reaction is dissipated rapidly from the thin, large-surface-area paint film and is subsumed, as it were, into the overall heat balance (see Chapter 4.1).

Organic chemical reactions can be divided into three categories in accordance with the nature of the reactive intermediates:

- polar (ionic, heterolysis)
- free-radical (homolysis)
- pericyclic

Comment on notation used for structural formulae

Organic molecules, and hence binder molecules, consist essentially of chain-like or ring-shaped backbones of hardly any chemical reactivity to which reactive functional groups of atoms are attached. The unreactive moieties in structural formulae are usually denoted by "R" or, as in binder chemistry, written as a wavy line if the moiety is long. For many purposes and for greater clarity, singly bonded fragments or moities are omitted altogether; instead, only the bonds to them are shown (as dashes). In modern notation, zig-zag lines denote hydrocarbon chains, with every change of direction and each free end marking the position of a carbon atom (C); again for clarity, it is standard practice to leave out hydrogen atoms (H) attached to carbon. The C and H atoms can be explicitly included in the zig-zag chains to highlight specific features. Naturally, hetero atoms, i.e. non-carbon atoms - along with other attached atoms - must be shown in the form of their chemical symbol, because otherwise they would not be identifiable as such. In this book, structural formulae are only written in zig-zag form when the formula would otherwise be confusing or too large.

5.5.2 Timing in organic reactions

All chemical reactions consist in the breaking of bonds in the **reactants** and the forming of new bonds to yield **products**. Most chemical reactions proceed via several individual steps called **elementary reactions**, which comprise bond breaking, bond formation or both to yield unstable intermediates or reactive intermediates. Elementary reactions are always reversible in principle and are considered irreversible if they essentially proceed from left to right only, i.e. the respective backward reaction is very slow. If both the forward and the backward reaction proceed at an appreciable rate, a chemical equilibrium is established between the various intermediates; in reaction equations, it is indicated by arrows pointing both ways. Chemical equilibria are quasi-stationary states in which the participants are striving to come to rest. However, when individual substances (reactants) are continually being formed and entering into further reactions, as in the case of reaction sequences, the equilibria are disturbed and are called flow equilibria.

As an example of a **reaction mechanism**, i.e. the sequence of elementary reactions constituting the overall reaction, consider the reaction of a reactant (R) to yield a principal product (P_p) and a minor product (P_M) via two intermediates (I_1, I_2):

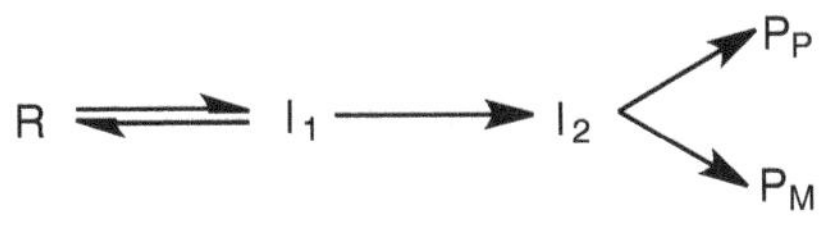

The first elementary reaction leads to **pre-equilibrium**; I_1 is formed rapidly (but not necessarily so) and its concentration is virtually constant. The next steps are irreversible, i.e. the equilibria are almost entirely on the product side, because the backward reactions are extremely slow. The step from I_1 to I_2 is often the slowest, i.e. the **rate-determining step**. The ratio in which P_P and P_M are formed is equal to the ratio of the reaction rates of the two **competing reactions** of I_2.

In each elementary reaction, all the molecules involved in that reaction step pass through an energy maximum, in which the molecules exist in the form of an **activated complex** or **transition state**. The height of the energy maximum for the reactants or products of the respective elementary reaction is called the **activation energy**, E_a, of the forward and backward reactions. Figure 5.15 shows the schematic energy profile of the reaction mechanism discussed above.

The general concepts needed for understanding the sequence of chemical reactions are as follows:

- the lower the activation energy of an elementary reaction, the faster it proceeds

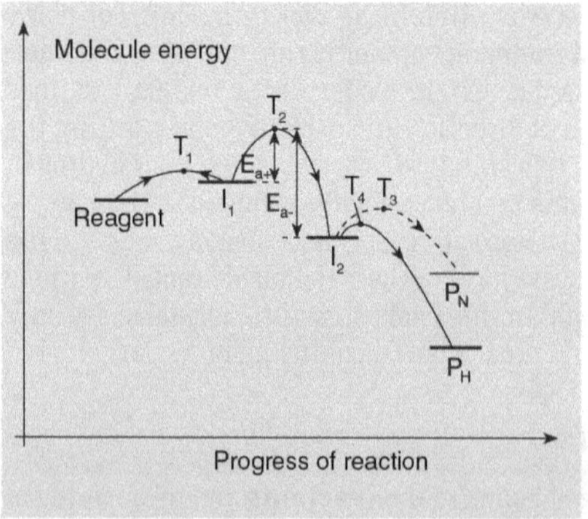

Figure 5.15: Energy profile of the reaction mechanism ($T_{1...4}$ are the transition states; E_{a+}, E_{a-} are the activation energy of the forward and backward reactions; the same applies to other elementary reactions) described in the text

- the forward and backward directions of an elementary reaction usually have different activation energies. The difference between the activation energies determines the position of the equilibrium of the elementary reaction
- the lower the activation energies involved, the faster the equilibrium is attained
- the slowest reaction of an elementary reaction mechanism (sequence of elementary reactions) determines the rate of the overall reaction (rate-determining step)

A further fundamental concept is that of **reaction molecularity**. This indicates the number of molecules that must be involved in the slowest, i.e. rate-determining, elementary reaction. A reaction is termed a mono (or uni), bi or tri molecular reaction according to whether a molecule reacts with itself or forms the activated complex with one or two other molecules.

5.5.3 Chemical kinetics

For the reaction

$$n_A A + n_B B + ... \longrightarrow n_C C + n_D D + ...$$

in which the reactants or reactive groups A, B etc, are converted to products or groups C, D etc, the reaction rate is given by

Equation 5.19: $$r = k_R [A]^\alpha [B]^\beta ...$$

[ ] is the concentration (in mol l^{-1})
α, β are exponents (usually integers, including zero)
k_R is the rate constant

The sum of the exponents $\alpha + \beta + ...$ is called the **order** of the reaction. The order is often approx. equal to the molecularity of the rate-determining elementary reaction of the overall reaction. As can be seen, the reaction rate increases with increase in concentration of the reactants or reactive groups.

During the reaction, the concentration of products increases. If the overall reaction is reversible, then, in accordance with Equation 5.19 (in which the products are the reactants for the backward reaction), the reaction rate of the backward reaction rises steadily until it matches the rate of the forward reaction, which is declining. When this happens, a dynamic equilibrium or **chemical equilibrium** exists, i.e. the concentrations of all substances or reactive groups are constant in time. This equilibrium is described by the **law of mass action** :

Equation 5.20: $$K = \frac{[C]^\gamma [D]^\delta ...}{[A]^\alpha [B]^\beta ...}$$

K is the equilibrium constant
α, β, ... are the individual orders of reaction for the forward reaction
γ, δ, ... are the individual orders of reaction for the backward reaction

The magnitude of the **equilibrium constant,** K, depends on the reactivity of all substances or groups (reactants and products) and on the temperature. (Pressure dependence is only of interest in gas reactions.) The temperature dependence can be readily described by the principle of least constraint (Le Chatelier and Braun):

- an increase in temperature shifts the equilibrium of an exothermic reaction to the left, i.e. towards the reactants, while the equilibrium of an endothermic reaction is shifted to the right, i.e. towards the products.

This can be observed with polymers: polymerisation is always an exothermic reaction. That is why strong heating of a polymer causes the polymerisation equilibrium, which lies far to the right (with the polymer) at room temperature, to move to the left, i.e. towards the monomers; the result is partial depolymerisation. The temperature at which this begins to an appreciable extent is the **ceiling temperature.**

From Equation 5.19, the reaction rate is determined directly by the **rate constant** k_R apart from the concentration of the reactants. As shown by **Arrhenius' equation,**

Equation 5.21:

$$k_R = A e^{-\frac{E_a}{RT}}$$

A is the pre-exponential factor
E_a is the molar activation energy
R is the general gas constant (8.314 J mol^{-1} K^{-1})
T is the absolute temperature in Kelvin

the former can be attributed to the activation energy of the reaction or the rate-determining elementary reaction. A is the rate constant at almost infinitely high temperature. To illustrate application of the equation, consider how temperature increases of 10 K from 300 K (27 °C) accelerates a reaction for three different activation energies:

$$\bullet\ E_a = 100\ \mathrm{kJmol^{-1}}:\ k_{310}/k_{300} = e^{\frac{100\cdot 10^3}{8.314}\left(\frac{1}{300}-\frac{1}{310}\right)} = 3.64$$

$$\bullet\ E_a = 60\ \mathrm{kJmol^{-1}}:\ k_{310}/k_{300} = 2.17$$

$$\bullet\ E_a = 20\ \mathrm{kJmol^{-1}}:\ k_{310}/k_{300} = 1.30$$

The results are consistent with the empirical observation that the reaction rate of most chemical reactions doubles to quadruples when the temperature increases by 10 K. It can also be seen that the temperature change has less influence as the activation energy decreases.

Interpretation of certain phenomena or observations in chemical reactions requires a further refinement to the theory. Eyring applied statistical thermodynamics to re-interpret Arrhenius' equation. The **Eyring equation** is:

Equation 5.22:

$$k_R = \frac{k_B T}{h} e^{\frac{\Delta S_m^{\neq}}{RT}} \cdot e^{-\frac{\Delta H_m^{\neq}}{RT}}$$

k_R is the rate constant
k_B is Boltzmann's constant ($8.381 \cdot 10^{-23}$ J K^{-1})
h is Planck's constant ($6.626 \cdot 10^{-34}$ J s)
ΔS_m is the molar activation entropy
ΔH_m is the molar activation enthalpy

Comparison with Arrhenius' equation (Equation 5.21), with allowance for the van't Hoff reaction isobars $[d(\ln K)/dT]_p = \Delta H_m/(RT^2)$ (see textbooks on physical chemistry) yields the following:

$$E_a = \Delta H_m^{\neq} + RT \approx \Delta H_m^{\neq} \qquad A = \frac{k_B T}{h} e^{\frac{\Delta S_m^{\neq}}{R} + 1}$$

This means that an elementary reaction will proceed rapidly when the **enthalpy of activation** is small and the **entropy of activation** is large. The latter is the increase in molecular disorder on proceeding from the reactants or reactant (of the elementary reaction) to the transition state. Since the transition state is more orderly than the free-moving reactant molecules, the entropy of activation is negative. A large entropy of activation, in the mathematical sense, is thus a slightly negative activation entropy, a fact which corresponds to only a small increase in order.

In other words:

Supposition: Activated complex is much more "orderly" than the starting molecules

$\Rightarrow$ strong loss of entropy during activation, i.e. strong negative activation entropy
$\Rightarrow$ small value for A
$\Rightarrow$ small value for rate constant k (slow reaction)
but: Strong acceleration of the reaction by temperature increase when E_a is large.

Example: For the purpose of good storage stability, 1-pack stoving paints may only react extremely slowly at room temperature, but must undergo rapid full cure at the lowest possible stoving temperature. This is achieved by a high value for A and E_a:

- a large value for A (due to a slight increase in order during activation) provides for a rapid reaction
- a large value for E_a renders the reaction rate heavily dependent on the temperature: at room temperature, the exponential term is so small that the reaction barely occurs at all (in spite of the large value for A). At elevated temperature, on the other hand, it is so large that the crosslinking reaction – supported by the large A – now proceeds rapidly

Unfortunately, the rate-determining steps of crosslinking reactions are usually bimolecular and thus tend to have low A values. However, the use of blocked crosslinkers or blocked catalysts can force the reactions in the direction of monomolecularity, with a resultant increase in A factors [22].

5.5.4 Polar reactions

In chemical reactions, bonds are broken and different ones are generally formed. This bond cleavage can proceed either **homolytically**, with formation of free-radicals (see Chapter 5.5.5), or **heterolytically**:

$$\overset{\delta(-)}{N} - \overset{\delta(+)}{E} \rightarrow NI^- + E^+$$

Heterolytic bond cleavage (heterolysis)

N-E is a polar molecule (or bond therein)
δ(-), δ(+) are partial charges

If the starting molecules carry a charge, the fragments can be uncharged. The partial charges indicate the bond's dipolar nature, which results from the different electronegativities of the directly bound adjacent atoms or from induction (see textbooks on organic chemistry).

The species with the free electron pair is a nucleophile, i.e. a particle searching for a nucleus, where "nucleus" means an atom (of a molecule) with a partial or full positive charge. The nucleophile - to some extent in a reversal of the cleavage reaction - then uses its electron pair to form a new bond to the positive reaction partner. The positive species, the **electrophile**, seeks a negative or electron-rich atom for the purpose of forming a new bond; i.e. it is electrophilic.

This is the base model by which all polar or ionic organic reactions proceed. However, the reaction mechanisms frequently entail complicated formulae due to the scope for charge shifts and electron shifts, often in combination with **mesomerism**. There are also several rules or principles that must be followed when reaction mechanisms are being interpreted or formulated (see textbooks on organic chemistry).

Polar reactions are the most common type of reaction in organic chemistry. They are favoured or triggered and possibly accelerated by

- strong polarity (polarized bonds) and/or alternatively
- good polarisability of bonds in the reactants
- polar solvents
- addition of ionic or ion-forming or highly electrophilic or nucleophilic additions (Lewis acids and bases) to the reaction mixture

The last-mentioned of these are **catalysts** if they are essentially not consumed during the reaction, and **accelerators** or **initiators** if they are consumed.

5.5.5 Free-radical reactions

Free-radical reactions frequently commence with **homolytic cleavage (homolysis)** of bonds:

$$R_1 - R_2 \longrightarrow R_1^\bullet + R_2^\bullet$$

Homolytic bond cleavage (homolysis)

$R_1^\bullet$, $R_2^\bullet$ are free-radicals

Homolysis may be triggered by high temperature, i.e. strong vibrations of bound atoms, or by capture of radiation. The latter may be electromagnetic (light, ultraviolet, gamma rays) or corpuscular in nature (beam of electrons). The resulting **free-radicals** are neutral or charged molecules (free-radical ions) that have at least one unpaired electron. They are generally highly energetic and reactive, because the electron would like to "pair". Examples of reactions by free-radicals are:

$$R^\bullet + A\text{-}B \longrightarrow R\text{-}A + B^\bullet$$
$$R^\bullet + A{=}B \longrightarrow R\text{-}A\text{-}B^\bullet$$
$$R^\bullet + S^\bullet \longrightarrow R\text{-}S$$

The secondary free-radicals generated in the first and second examples quickly enter into further reactions, as a result of which chain reactions often occur. The most important example of this reaction is **free-radical polymerisation**, which is also a hugely important curing reaction (see Chapter 10). The third example constitutes **free-radical combination** in which the excitation energy of both free-radicals is lost and the chain reaction stops.

An important, relatively low-energy but ubiquitous free radical which can trigger or terminate free-radical reactions – as a function of reaction conditions and reaction partners – is the oxygen molecule, O_2. In its electronic ground state, it is a **diradical**: •OO•.

Free-radical reaction with oxygen is the mechanism by which oils, alkyd resins and other film-formers undergo **oxidative drying** (crosslinking, see Chapter 7).

Free-radical reactions are favoured, or triggered by

- good electronic stabilisation of the free-radical state (mesomerism)
- high temperatures (greater than 250 °C)
- short-wave radiation (UV, blue light)
- nonpolar solvents
- attacking radicals or free-radical formers (free-radical initiators)

though not all conditions need to be met simultaneously.

5.5.6 Pericyclic reactions

This type of reaction plays hardly any role in the chemistry of film formation. However, it could prove attractive for future developments in this area and so a brief explanation is provided below.

At elevated temperature or given photochemical excitation, certain molecules having one or more double bonds can readily enter into intramolecular or intermolecular reactions, in which a bond shift (electron pair shift) occurs in an elementary step via a cyclic transition state. Since neither free-radical nor ionic intermediates can be detected, it would appear that a different reaction principle is involved. The driving force here is the transition of high-energy π bonds to low-energy σ bonds[16], i.e. from double bonds to single bonds. In paint chemistry, the **Diels-Alder reaction**, also known as the diene synthesis or [4 + 2]-cycloaddition, is of primary importance. It consists in the reaction between the double bond of one molecule, the **dienophile**, with two conjugated double bonds of a second molecule, the **diene**, at elevated temperature to form a six-member ring, e.g.

Examples of Diels-Alder reactions in paint chemistry are the reaction between unsaturated binders and maleic anhydride, the dimerisation of unsaturated fatty acids or fatty acid chains, and the addition reaction of drying oils to resols (via quinone methides).

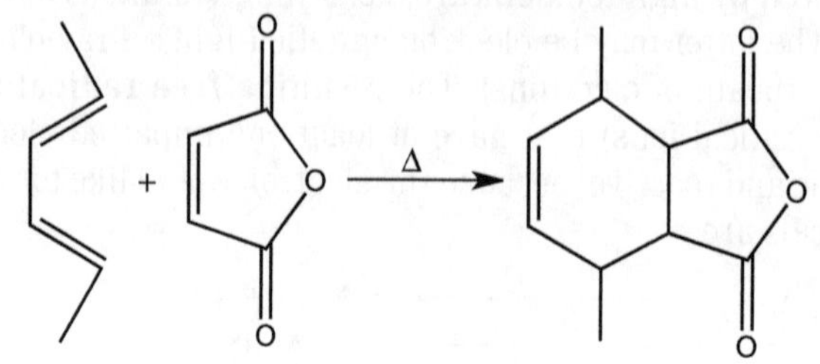

Diene (e.g. fatty acid chain) | Dienophile (here: maleic anhydride) | Diels-Alder adduct

[16] A double bond consists of a stable (low-energy) σ bond and a weaker (high-energy) π bond

6 Physical drying

The influence of solvent evaporation and film thickening on wet-film formation was already covered in Chapters 3.2 to 3.4 while Chapters 4.1 and 4.2 dealt with the physical aspects of drying, particularly concerning heat input into the paint layer or the object for painting. Chapter 5.2 provided extensive basic information about solvents and polymer solutions or binders. The reader is especially reminded of the WLF equation (Equation 5.7) and Figure 5.8 in this regard. What follows now is to be seen in the context of those sections, and applies not only to full physical drying of films, but partly also to initial surface drying prior to curing.

6.1 Physical drying from solutions

6.1.1 Solvent transfer from film to ambient air, and heat balance

Equation 4.1 shows the general heat-transport equation for heat transfer. All the specific information about the process concerned is tucked away inside the heat transfer coefficient, α. An entirely analogous equation can be formulated for mass transfer - for our purposes, this means the flow of solvent vapour from the wet film surface into the interior of the adjacent gas phase (ambient air, dryer air):

Equation 6.1: $$\dot{n}_D = \beta_n A (c_{D,F} - c_{D,L})$$

$\dot{n}_D$ is the mole flux across surface A
β_n is the mass transfer coefficient (in molar form)
$c_{D,F}$ is the vapour concentration (n/V) at the film surface
$c_{D,L}$ is the vapour concentration inside the ambient air

If the solvent vapour is considered to be an ideal gas, p = c RT can be used to substitute the pressure for the concentration:

Equation 6.2: $$\dot{n} = \beta_n A \frac{p_{D,F} - p_{D,L}}{RT}$$

$p_{D,F}$ is the solvent vapour pressure at the film surface
$p_{D,L}$ is the solvent vapour pressure inside the ambient air

The solvent vapour pressure in the ambient or dryer air must be kept low for rapid evaporation; in the open air, it is practically zero for organic solvents. The water vapour pressure in the air, according to $p = p_s \phi / 100\,\%$ (where ϕ is the relative humidity in %, p_s is the saturation vapour pressure), is finite however, and exerts a strong influence on the evaporation rate. The damper the air, the more slowly the water evaporates.

Since every mole of solvent that evaporates requires input of the molar evaporation enthalpy, $\Delta_v H_m$, the following holds for the necessary evaporative heat flow $\dot{Q}$

Peter Mischke: Film Formation in Modern Paint Systems

ISBN 978-3-86630-861-9

Equation 6.3:

$$\dot{Q} = \dot{n}\Delta_v H_m = \beta_n A \frac{p_{D,F} - p_{D,L}}{RT} \Delta_v H_m$$

Let us assume a quasi-steady state exists, i.e., all parameters are virtually constant in time, and that this heat flow is introduced into the surface film only by heat transfer from the air, in accordance with Equation 4.1:

$$\dot{Q} = \alpha A (\vartheta_L - \vartheta_F)$$

ϑ_L is the temperature inside the ambient air
ϑ_F is the temperature of the film surface

Combining this equation with equations 6.2. and 6.3, we get:

Equation 6.4:

$$\frac{\dot{n}}{A} = \frac{\alpha}{\Delta_v H_m}(\vartheta_L - \vartheta_F)$$

This states that the solvent vapour flow (mole flux density) per unit area, $\dot{n}/A$, increases with increase in heat transfer coefficient, α, and with increase in the temperature difference between the ambient air and the wet film surface, and is inversely proportional to the enthalpy of vaporisation $\Delta_V H_m$. Again, water exhibits anomalous behaviour: its enthalpy of vaporisation is a multiple of that of organic solvents, which is the major reason for its high evaporation number of 80 compared to, e.g., Figure 6.1 of toluene (bp 111 °C).

As described in Chapter 4.1, α increases with increase in the air-flow rate. Comparison of Equation 6.1 with Fick's first law (Equation 5.17) shows further that β_n is formally equivalent to $D/\Delta x$, i.e., is inversely proportional to the diffusion layer thickness, Δx. However, the latter falls with increase in flow rate, a fact which boosts the vapour-diffusion flow, i.e., also leads to accelerated drying.

These results may be paraphrased as follows:

- the rate of physical drying increases with
- increase in air-flow rate
- increase in the temperature difference between the ambient air and the film surface
- increase in the difference between the solvent and water vapour pressure in the ambient air and at the film surface
- decrease in the enthalpy of vaporisation or increase in volatility of the solvent.

It must be emphasised, however, that these variables, especially the thermal ones, are interdependent, i.e., affect each other, in complex ways. For example, a lower enthalpy of vaporisation leads to a higher vapour pressure or greater volatility, causing the film surface to cool more extensively, especially when the air-flow is pronounced; this lowers the vapour pressure and retards evaporation, but increases the rate of heat transfer, etc. The overall result depends heavily on the predominating effects.

Purely experimentally, it can be observed that blowing off highly volatile liquids leads to marked cooling, which is called **evaporative cooling** in the industry. Initially very rapid, evaporation slows itself down, as it were, by cooling the surface. Consequently, the difference between the film and air temperatures increases, heat input increases and the process approaches a steady state.

This cooling of the surface can lead to blushing, which is a negative effect. **Blushing** occurs when the surfaces of paint films that dry very quickly – particularly those based on cellulose nitrate and low boilers – cool below the dew point of the air, giving rise to condensation and encapsulation of fine water droplets in the film surface. When the water evaporates, it leaves behind small voids in the film surface that cause white light scattering. Blushing can be prevented by optimising the drying parameters and the solvent composition. A particularly effective solution is to add polar, moderately volatile and relatively hygroscopic solvents, such as butanol, which initially boost compatibility of the paint with water and then quickly entrain the latter from the film by forming an azeotrope [17, 61]. Aromatics, such as xylene and toluene, have good entraining action, too, but are generally undesirable for toxicological reasons.

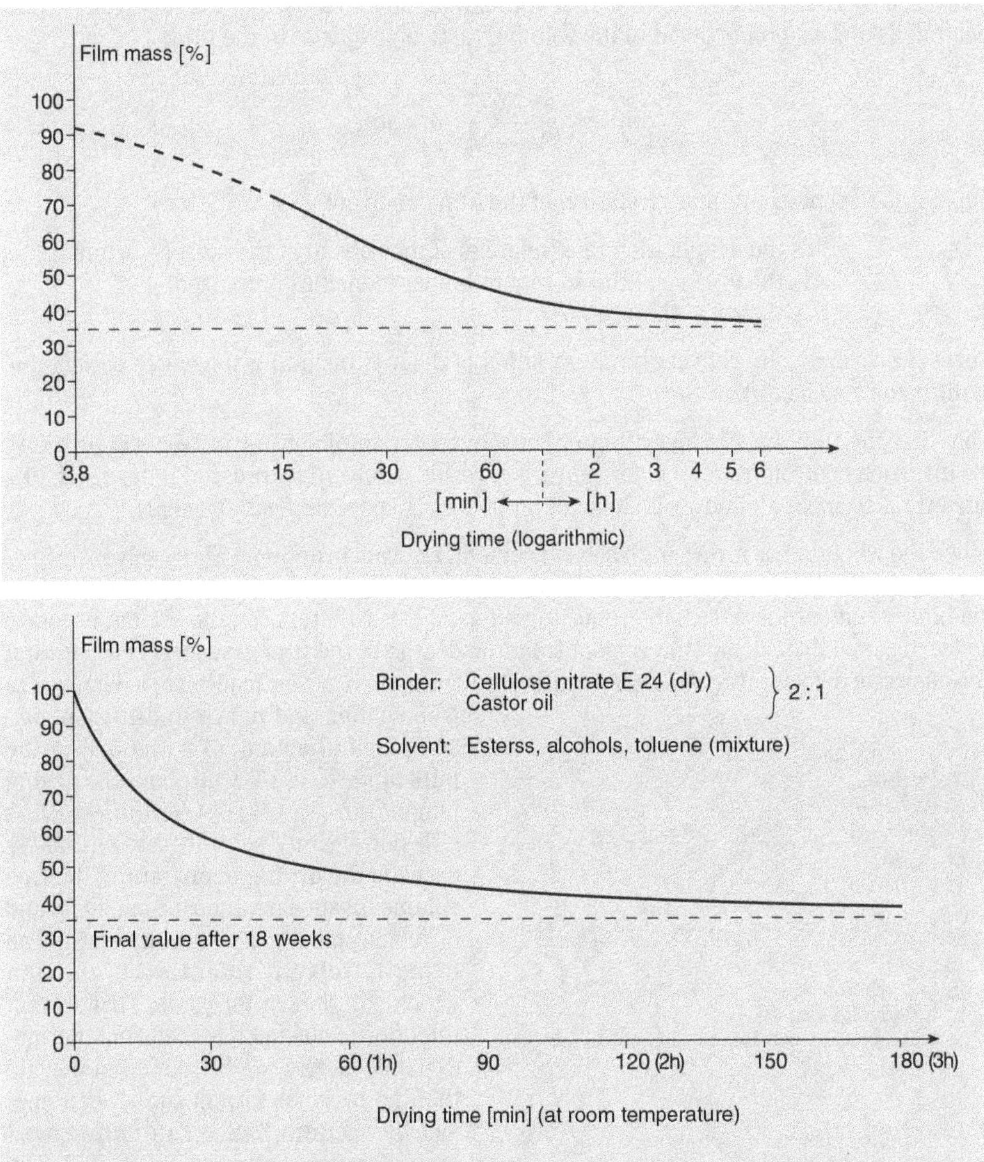

Figure 6.1: Drying of a cellulose nitrate clearcoat over time. a) linear time-scale b) logarithmic time-scale (adapted from [62])

6.1.2 Sequence of events during drying

6.1.2.1 Evaporation, diffusion and solvent retention

Figure 6.1 (a and b) shows the experimentally determined loss of mass undergone by a physically drying clearcoat film over time [62]. The logarithmic plot (see Figure 6.1b) clearly shows the asymptotic change in drying over long period times. Drying took 18 weeks to complete. This is evident from the fact that heating the film at 100 °C yielded no further loss of mass; permanent solvent retention had therefore not occurred (see below).

Figure 6.2 compares three theoretical drying curves obtained under different assumptions. Curve 1 represents the decrease over time in the mass of a 20 % binder solution, with the evaporation rate, expressed in terms of film mass, set at a constant (negative) 1 % min^{-1}. For curve 2, it was assumed by way of a first approximation to reality that the relative evaporation rate is always proportional to the volume fraction of solvent in the film:

$$dm = \dot{m}_{S,0} \frac{V_S}{V_S + V_B} dt = dm_S$$

dm, $dm_S$is the change in mass (relative) of the film or solvent

$\dot{m}_{LM,0}$	is the evaporation rate (relative) of the solvent at the start (-1 % min^{-1})
$V_{S,}$ V_B	is the volume of the solvent or binder (densities equalized)
dt	is the time differential

Curve 3 schematically reflects the retardation of drying, including solvent retention, due to diffusion (see below).

Note that the time axis is logarithmic. Thus, evaporation of solvent in curve Figure 6.1b (no diffusion control) from 10 g down to 1 g of residual solvent (expressed in terms of 20 g binder) takes about an hour, which is not immediately apparent from the chart.

When the wet film has a very high solvent content, i.e., well over 50 wt.%, the solvent evaporates almost as quickly as the pure solvent, i.e., as indicated by the evaporation number. As the binder content increases, the retarding effect of this “dilution” on the solvent becomes noticeable, and diffusion of the solvent from the depths of the film gradually slows until it becomes rate determining. Because the film is drying, the T_g rises, leading to a decrease in free volume and hence in diffusion constant (see Chapters 5.1.8 and 5.4). If the pure binder has a T_g well above the drying temperature, the T_g of the film, which is still not entirely solvent-free, ultimately exceeds the drying temperature; the free volume disappears almost completely and diffusion comes to a virtual standstill. The result is **solvent retention** in the film which can persist for years. That solvent retention is due to kinetic and not thermodynamic factors can be seen from the fact that the residual solvent cannot be removed by vacuum, but it can be removed by prolonged heating above T_g [22]. The residual solvent cannot be detected as any appreciable softness or tackiness of the

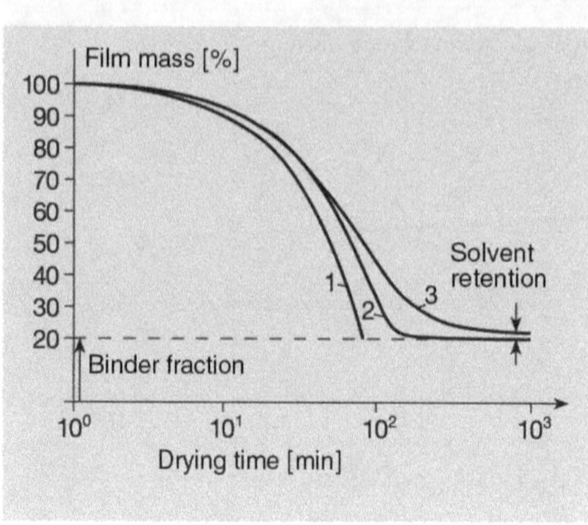

Figure 6.2: Drying curves for wet films as per simple computational models (1 and 2) and under diffusion inhibition (3) (see text for explanations)

film, of course, since the film is necessarily vitrified. Nevertheless, solvent retention weakens the film and carries the risk of persistent, low-level solvent emission, which is very undesirable or inadmissible indoors or in the food sector.

The dependence of the diffusion rate of solvents on the free volume, i.e., the number and size of thermally generated voids in the polymeric binder matrix, can be explained by the fact that the principal diffusion mechanism entails solvent molecules "jumping" from hole to hole. Small solvent molecules tend to leave the film faster than large ones for two reasons:

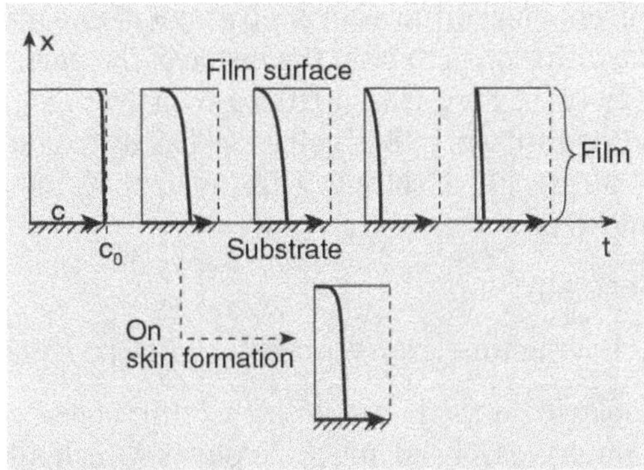

Figure 6.3: Solvent concentration profiles versus drying time

- first, volatility increases with decrease in molecular size, and
- second, the small molecules fit into more holes, as a result of which their "jump probability" increases [22]

Cyclohexane is almost twice as volatile as toluene, but persists to a greater extent in the film since its bulky molecules do not fit as readily into the holes as the flat toluene molecules. Similar observations have been made for solutions of acrylic resin or cellulose nitrate in a mixture of iso- and n-butyl acetate (BAC and IBAc) 60:40. The more volatile IBAc initially evaporates faster from a wet film containing high solvent excess. From a solvent concentration of 20 wt.%, expressed in terms of the binder, the IBAc fraction in the remaining solvent then rises again (to approx. 90 %), since its bulky molecules diffuse more slowly than the molecules of does BAc [22].

Although physical drying under real conditions is not entirely amenable to mathematics, it is possible to gain an idea of the change in solvent concentration in the paint layer over time. Paint films dry from the outside in; thus the T_g of the film will decrease from the surface to the substrate, while the opposite applies to the diffusion coefficient. A small diffusion coefficient at the surface retards post-diffusion of solvent from the deeper layers, a fact which, according to Fick's first law, leads to flattening of the concentration gradient in deep layers. In the extreme case, over-rapid surface drying causes **skinning** or vitrification of the surface, and brings drying in the deeper layers almost completely to a standstill. Over-rapid crosslinking of the surface during oxidative drying or curing by stoving can also produce a barrier effect of this kind. Figure 6.3 shows concentration profiles which have been sketched by the author.

6.1.2.2 Influence of layer thickness

There is a saying in the painting trade, where physical drying can still be an important time factor, that two thin coats are better than a thick one [7]. This rule can easily be explained by a thought experiment:

As explained in Chapter 6.1.2.1, the drying rate for the relatively high solids contents (nonvolatiles fraction) of paints at the advanced stage is mainly diffusion controlled. Now imagine two wet coats, which differ only in that one is twice as thick as the other. Let one coat (A) be 50 μm thick and the other (B), 100 μm. In this case, the 1-μm-thick partial layer (PA) at a depth of 10 to 11 μm in A corresponds to the 2-μm-thick partial layer (PB) at 20 to 22 μm in B. To achieve the same degree of drying in these partial layers, twice as much solvent must diffuse out of PB as out of PA, and that would also take twice as long. However, the diffusion path from the inside of the partial layer to the surface in B (21 μm on average) is twice as long as in A (10.5 μm on average). In accordance with Fick's first law (Equation 5.17), diffusion in B is half as fast as diffusion in A, i.e. diffusion takes twice as long for the same amount of solvent. This quadruples the total drying period for a doubling of the layer thickness.

Generally:

- Drying time grows (roughly) with the square of the layer thickness

Replacing a thick coat with two thin ones has another advantage. Generally, dry coats of paint have isolated, ultrafine **pores**, which allow migration of corrosive media through the coat. If two coats are applied one on top of the other, the probability that a pore of the first coat will coincide with a pore of the second coat to yield a common pore is practically nil. This technological advantage must be offset against the higher labour costs for multi-layer coatings, of course. Furthermore, when previously applied layers are over-painted, reduced **interlayer adhesion** or **lifting** may occur. In the latter case, the fresh wet film dissolves the underlying coats extensively.

6.1.3 Solvent selection

Perfect physical drying or surface drying (in crosslinking paints) requires mixtures of two or more solvents. Often, the solvent mixture is composed in such a way that its solubility parameter (point) is located at the edge of solubility region of the binder (see Chapter 5.2.2.2). During drying, however, the solvent composition should change to allow the resulting parameter to migrate to the centre of the solubility region, i.e., to the binder parameter. The non-dissolving or poorly dissolving solvent (non-solvent, diluent) must therefore be more volatile than the good solvent, because the latter must build up during drying. Figure 6.4 illustrates the principle for a binary solvent mixture.

For air-drying paints that dry at room temperature, the following rule of thumb applies to solvent selection:

- approx. 45 % low boilers
- approx. 45 % moderate boilers
- approx. 10 % high boilers (good solvents)

Low and perhaps moderate boilers should not be used for stoving paints, since excessively fast evaporation or evaporation of solvent fractions may lead vapour bubbles (**popping**), which leave craters behind because the viscosity rises rapidly due to the onset of curing [54].

It should also be recalled that thermodynamically good solvents have a high affinity for binders, which – in combination with the poor volatility – may boost the risk of solvent retention (see Chapter 5.2.2.3).

If the solvent mixture which remains is a good solvent (which it is required to be), it promotes levelling and thus formation of gloss by the paint; however, if it is a poor solvent or even a non-solvent, levelling problems can be expected due to the strong rise in viscosity and precipitation of the binder. The molecular film structure, too, is determined by the solvating power of the solvent. Good solvents lead to greater widening of coils or uncoiling of the binder molecules than poor ones. This promotes coil interpenetration and entanglement, and the dried film has the network structure shown in Figure 5.4; the result is a high film strength. Poor solvents can shrink the coils on the other hand, which are then penetrated

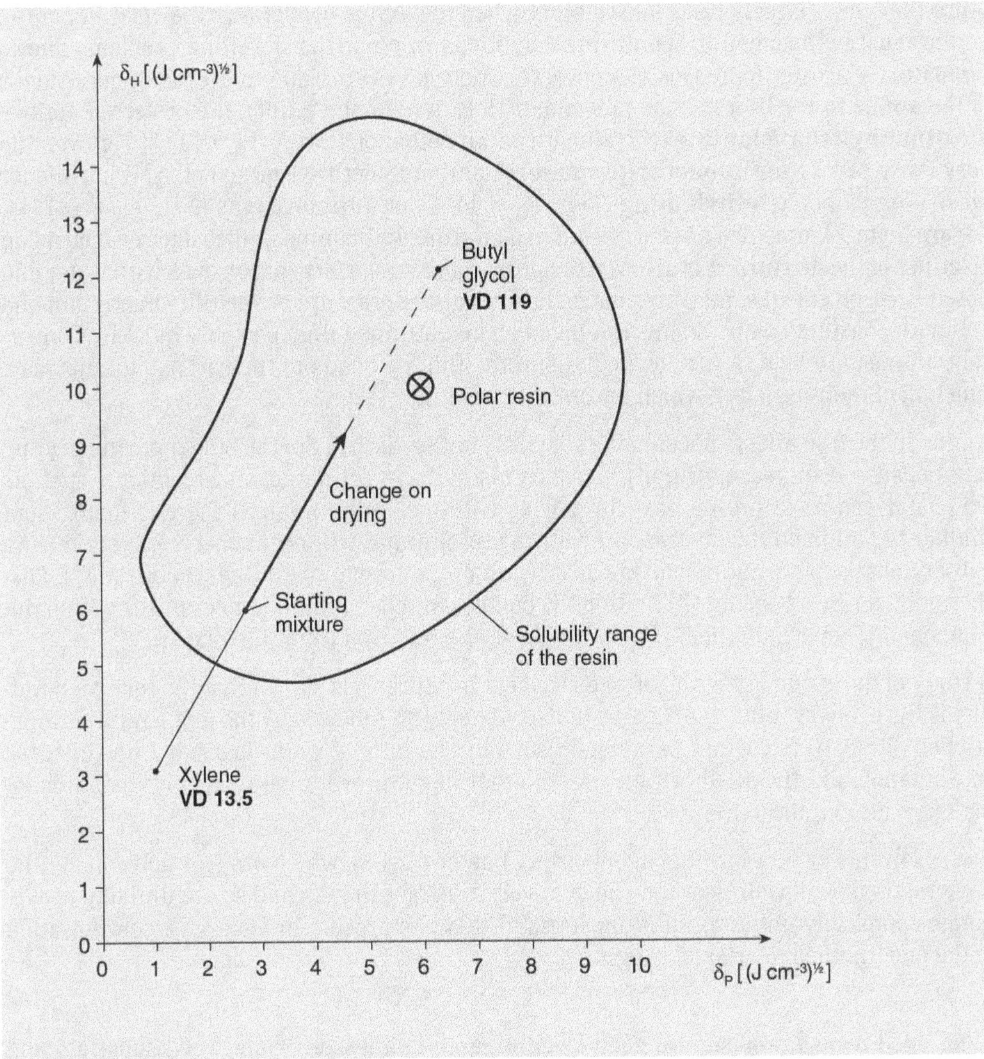

Figure 6.4: Advantageous shift in solubility parameter of a solvent mixture (xylene / butyl glycol approx. 2:1) relative to the binder parameter during drying

to a lesser extent in the wet film and subsequently in the dry film, i.e., tend to have the cell structure shown in Figure 5.4 [6].

6.1.4 Evaporation processes in aqueous systems

Unlike dispersions, true or colloidal binder solutions (see Chapter 5.3.2) exhibit weak physical surface drying at best – owing to the low molecular weights of the dissolved molecules. Stable films are only achieved by chemical crosslinking. However, the physical evaporation processes at the start of film formation exert a significant influence on the quality of the resultant coating.

In general:

- water, amine and perhaps co-solvents evaporate at different rates their volatilities need to be matched to the drying conditions (or vice versa)

Since the binder largely loses its solubility when the amine evaporates, the volatility of the amine must be matched to the drying conditions in order that levelling problems due to premature viscosity increases or, conversely, delayed drying due to over-long persistence of the amine in the film may be prevented. Thus, for stoving paints, the preferred amines are **dimethylethanolamine** (N,N-dimethylaminoethanol, DMEA, T_{bp} = 134 °C) or even the very low volatility **aminomethylpropanol** (2-amino-2-methyl-1-propanol, AMP); while for air-drying paints, **triethylamine** (TEA, T_{bp} = 86 °C) or **ammonia gas** (NH_3, gaseous) [14, 63] are better. Ammonia gas is also used in a mixture with amines, since it lowers the mean volatility of the mixture. It is also worth noting that amines form azeotropes with water and possibly cosolvents (i.e. mixtures which have a vapour pressure or volatility maximum and boil with constant composition), and thus can be entrained unexpectedly quickly. Conversely, amines are held to various degrees in the film by bonds to the carboxy groups (as a function of their basicity), which lowers their volatility [14].

Amine in the film affects not only the solubility or the viscosity of the binder during drying, but, because it increases the pH, perhaps also affects pH-dependent reactions, such as acid-catalysed crosslinking of resin polyols with melamine resin. If there is insufficient inhibitory amine on the surface, premature crosslinking will occur there, leading during subsequent deep curing to matting as a result of the formation of fine wrinkles [22]. Furthermore, amines bearing OH and/or NH bonds can react at high temperatures with paint components, especially the binder, and thereby be removed from the mixture.

A study of the evaporation rate of water/solvent mixtures has shown that the relative humidity is the primary influence (range studied: 10 to 90 %), followed by the temperature (range studied: 20 to 40 °C). Other parameters such as the ratio of water to solvent (in this case sec.-butanol) and the air-flow rate, on the other hand, prove to be relatively insignificant [4][17)] (see also Figure 2.13).

The cosolvents or film-forming agents (see Chapter 6.2.1.3), which are true solvents for the binders, promote levelling at the end of physical surface drying and so are unlikely to evaporate completely until crosslinking starts. If they are present in too high a concentration at the end, there is the risk of sagging.

In general:

- for good non-sag resistance, good levelling and, to a lesser extent, good substrate and pigment wetting, and gloss development, the water/solvent ratio during surface drying and drying must always lie in the optimal range [4].

Undesirable rapid or strong drop in viscosity or both may also occur if the binder concentration in the paint is still in the region of the water mountain when evaporation commences (see Chapter 5.3.2, Figure 5.13). When the water evaporates the concentration of binder increases and so moves into the region of lower viscosity. Modern waterborne coatings based on colloidal solutions and/or dispersions have virtually no water mountain, however.

The formation of an azeotrope from butyl glycol and water was explained in Chapter 5.2.1.2. As mentioned there, the composition of the azeotrope is temperature-dependent. Below the boiling point of the mixture at which the evaporation process is diffusion-controlled, the relative humidity exerts an added influence. Since the evaporation rate of butyl glycol to a first approximation is independent of the air humidity, while that of water drops with increase in humidity (see Chapter 6.1.1), a rise in humidity is accompanied by a shift in the composition of the azeotrope – i.e., without change in composition of the evaporating mixture – towards a higher water concentration. Conversely, for each solvent/water mixture,

[17)] Unfortunately the results here are confusing or greatly lacking in detail.

there is a so-called **critical relative humidity** (CRH), at which a mixture forms azeotropes. The CRH is dependent not only on the composition (including the binder) but also on the temperature.

Another rule of thumb is:

- CRH and relative humidity of the drying air should be as similar as possible, so that the water/cosolvent ratio does not change by too much

A special effect occurs when IR bright emitters with strong emissions at approx. 3 µm are used during the IR pre-drying or additional drying of waterborne paints that is frequently encountered in industry: water absorbs this radiation very strongly and is therefore expelled almost selectively, whereas the cosolvents evaporate relatively slowly [37].

6.1.5 Film formers for physically drying paints

Basically, a distinction must be drawn here between coalescing agents that dry from solution and dispersions, which are usually aqueous. Drying of dispersions is covered in the next section. This section will cover only a few important film formers that form finished paint films from solution. Thermoplastic film formers have higher average molecular weights, usually in the range of 20 to 50 $\cdot$ 10^3 g mol^{-1} (or higher), which lead to paints of high viscosity or low-solids content; the outcome is usually "thin" coatings with little **filling power (build)** and a high VOC level. Nonetheless, these binders retain some limited significance for one-component coating materials that are processed at and dry rapidly at room temperature. Moreover, the thermoplastic film formers have some general and specific technical characteristics that can sometimes be desirable. Since the coatings are not crosslinked, they will re-flow after being sanded and heated (reflow paints) or can be easily stripped. Given sufficient compatibility, the thermoplastics are frequently combined with other film formers that crosslink too, often in order to accelerate the physical surface drying or - e.g. in the form of low molecular **soft resins** - to confer flexibility. It must be borne in mind, however, that these film formers incorporate their thermoplasticity into the coatings, e.g. in the form of reduced solvent resistance. With favourable film formers, such as cellulose derivatives, the purely physically-drying film formers react with the crosslinker component of the paint and thus largely lose their thermoplasticity.

Examples of physically drying film formers:

- chlorinated rubber
- cyclo rubber
- vinyl chloride copolymers
- cellulose nitrate (nitrocellulose)
- cellulose acetobutyrate (CAB)
- thermoplastic polyacrylates
- polyvinyl esters (copolymers)
- hydrocarbon resins
- high-molecular epoxy resins (phenoxy resins)
- novolaks (unreactive phenolic resins)
- silicone resins
- polyamides
- bitumen, natural asphalt

6.2 Physical film formation from dispersions

6.2.1 Waterborne primary dispersions

6.2.1.1 Qualitative aspects of film formation

A general description of primary dispersions, i.e. emulsion polymers, was provided in Chapter 5.3.3. Film formation by this important type of binder and derived dispersion coating materials, such as emulsion paints and coatings, proceeds by a mechanism which is totally different from the physical drying of dissolved film formers. Evaporation, mechanical and diffusion processes all overlap at various time intervals. Several theories seek to explain these but largely arrive at different results and are sometimes contradictory. In light of this, the following approach will be taken here: Figures 6.5 to 6.7 will be used to provide a schematic first impression of the film-forming process. That will be followed by a description, supported by experimental facts. Chapter 6.2.1.2 will then briefly discuss some physical considerations and models.

Figure 6.8 charts the release of water from model dispersion over time. There are three stages. In section I, water release is rapid and proportional to the elapsed time. The evaporation rate is approx. 85 % of that of pure water. The particles are distinct from each other. In section II, the evaporation rate drops off quickly. The particles join together, lowering the surface area available for evaporation. In section III, water release proceeds very slowly, and essentially takes the form of diffusion. This is known as the **three-stage model of *Vanderhoff*** [65].

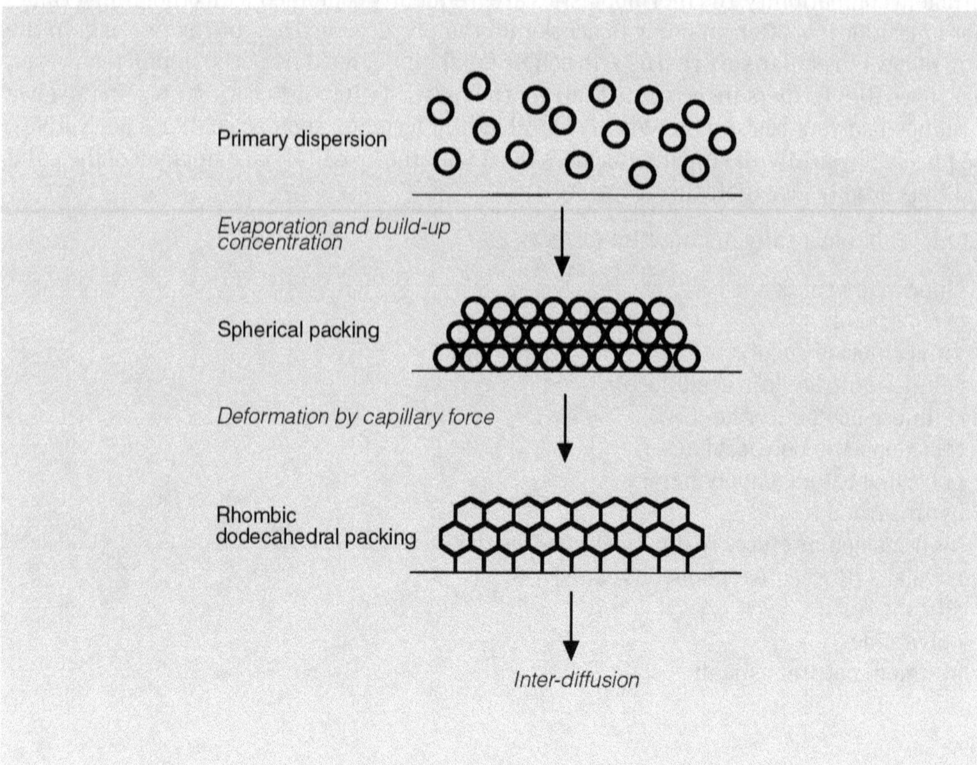

Figure 6.5: Principle behind film formation by dispersions; from [14]

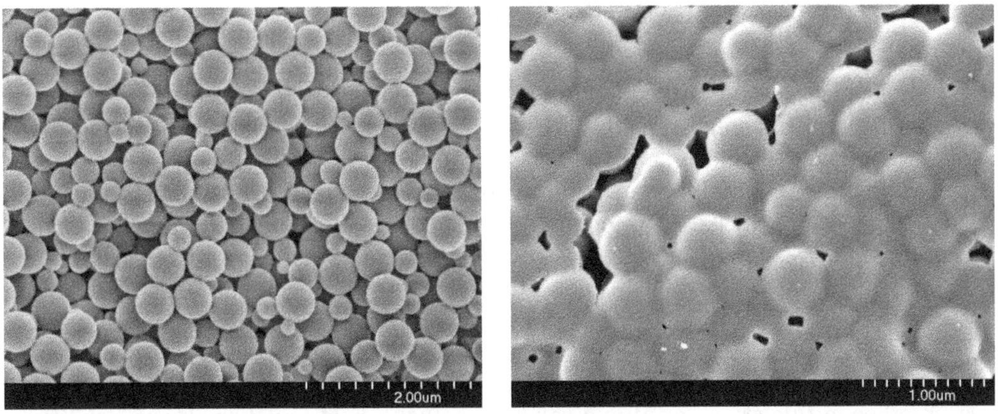

Figure 6.6: left: Dipersion before coalescence; right: dispersion widely coalesces; source: BASF SE [64]

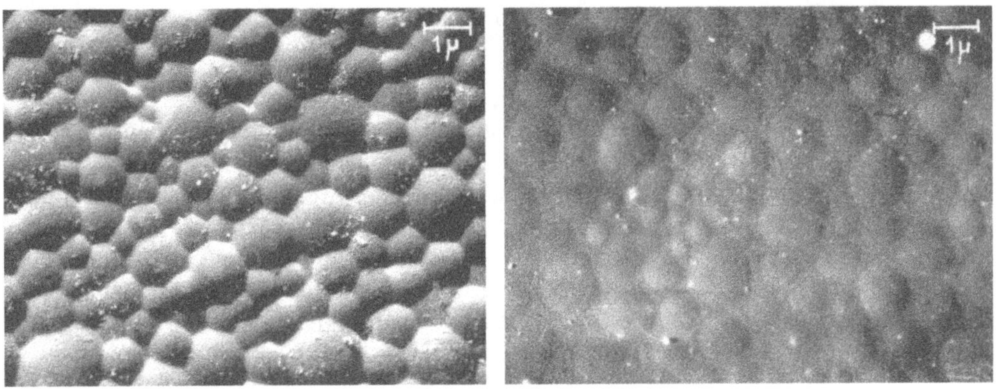

Figure 6.7: Scanning electron micrographs of film-formation by a dispersion, from [7]

On the basis of concurring descriptions in the literature [3, 10, 66], a more detailed description of the film-forming process may be as follows:

- after application of the dispersion, water evaporates from the surface of the wet film in a relatively slow process because of water's relatively high evaporation number of 80
- at a volume fraction of approx. 52 %, the particles are so close together that they touch, and **capillary forces** of attraction develop; the viscosity rises steeply in line with the Mooney equation (Equation 5.16)
- the spheres become more and more densely packed until, in the ideal or theoretical case the densest sphere packing (for equal-sized particles) is obtained at 74 vol.%
- further loss of water deforms the particles into rhombic dodecahedra having hexagonal symmetry. The shells of emulsifiers and possibly protective colloids are penetrated, leading to coalescence and **interdiffusion** of the polymer molecules across the original particle boundaries. The last bits of water are forced out under high pressure through the narrow capillaries, which form the residual gussets between the deformed particles

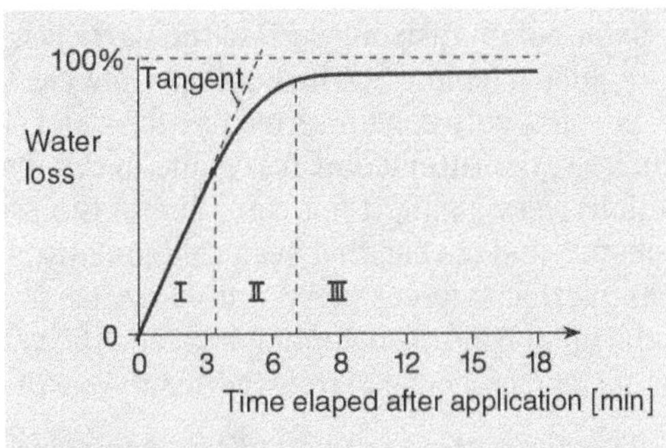

Figure 6.8: Drying curve of a poly(n-butyl methacrylate) dispersion at 40 °C; see text for explanation; adapted from [65]

This description is borne out by film-forming experiments:

- the rate of film formation is determined by the rate of water evaporation. If evaporation is inhibited, e.g., by very high humidity, film formation also slows down
- plastic dispersions composed of moderately crosslinked particles form a continuous film, too. Although crosslinking largely prevents mutual penetration of the particles, forces must be acting that are high enough to overcome the forces of the polymer/water interfaces and to cause the particles to fuse
- film formation by dispersion particles in the absence of water requires a much higher temperature than when it is present. This demonstrates the importance of water for coalescence [67]
- *Hwa* showed that, prior to reaching the densest spherical packing, many polymer dispersions already no longer exist as a liquid system, but are flocculated. The particles touch each other in this state, forming a coherent network held together by Van der Waals forces which are strong enough to prevent redispersion. *Hwa* observed this phenomenon in polymer volume fractions of about 0.5 to 0.6; in other words, even before the densest spherical packing was reached [68]

Primary dispersions are generally fast-drying binders. This has its advantages, to be sure, but it also has some drawbacks such as premature surface drying in the form of encrustation, skinning or stringing during processing and too short an **open time** after application. These difficulties can usually be adequately countered, however, by adjusting the formulation, e.g. incorporating additives or blending with truly dissolved binders to yield **hybrid systems** and/or by optimising the application conditions (higher humidity, lower application temperature). Interior emulsion paints, for example, contain cellulose ethers, for **water retention** purposes. They thus slow down drying to some extent, because, in line with what was said above, the rate of film formation depends on the rate of water evaporation.

6.2.1.2 Physical models

Brown assumed that at least four driving forces lay behind film formation, namely the

- force F_0 due to the surface tension of the particles and the desire to reduce the surface area
- capillary forces FC due to the surface tension of the aqueous phase of the menisci in the capillaries
- Van der Waals forces Fv between the particles
- gravitational force FG which causes the particles to settle [67]

These forces are opposed by the

- deformation resistance FD of the particles and
- repulsion force, FS, which arises from electrostatic and/or steric stabilisation [67]

F_0 only acts between unwetted particles and should thus be irrelevant. In the case of wetted particles, the **interfacial force** due to the interfacial tension $\sigma_{s,l}$ would act instead; this is initially very small, if not zero, due to the presence of surfactants. It might be concluded from this and the fact that even well-stabilised dispersions form films without any problems that interfacial forces play hardly any role at all. But it is precisely this assumption which is the main bone of contention between the various experts and theories [22, 68]. Two arguments lend credence to the significant role of interfacial forces:

- first, the entire process of film-formation must be thermodynamically driven, and this force can only come from the release of interfacial energy by coalescence

- second, the aforementioned good wetting or stabilisation of the particles only occurs in the undisturbed initial state of the dispersion; removing the water from the film leads to destabilisation, i.e., an increase in the interfacial energy

While interfacial forces are not mentioned explicitly in the following account, they could be included as an additional contribution to the capillary forces, without any qualitative change in the overall discussion. Depending on which model is selected for the shape of the voids between the polymer particles, the pressures generated by the interfacial forces would range from $\sigma_{s,l} / r$ to $2\ \sigma_{s,l} / r$ (r = radius of sphere).

The presence or relevance of capillary forces (F_C) is accepted throughout the literature. Figure 6.9 schematically illustrates the genesis of capillary forces or pressures. The (apparent) pressure under which a liquid having a surface tension of (e.g.) $30 \cdot 10^{-3}\ N\ m^{-1}$ in conditions of complete wetting moves into a capillary of 0.01 μm diameter (about 1/10 particle diameter) is calculated from the equation $p = 2\ \sigma / r$ 120 bar. This pressure propagates across the liquid column somewhat in the manner of suction, drawing the particles towards each other with a resultant force.

The strength of the inter-particle Van der Waals forces can be quickly estimated by a rough calculation. For spheres whose radius is larger than the distance between them, the following applies:

Equation 6.5:

$$V_v = -\frac{A_H r}{12H} \Rightarrow F_v = \frac{dV_v}{dH} = \frac{A_H r}{12H^2}$$

V_V is the potential Van der Waals energy of attraction
A_H is the effective Hamaker constant
r is the radius of the sphere
H is the distance between the spheres
F_V is the Van der Waals forces

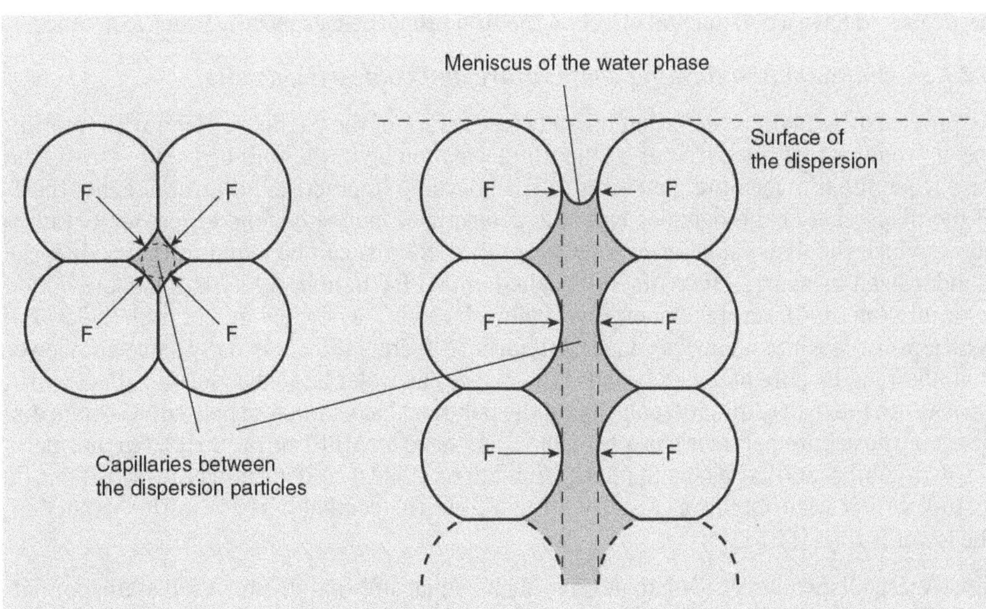

Figure 6.9: Genesis of capillary forces (schematic)

Assuming that A_H has a mean value of $1 \cdot 10^{-20}$ J [69], then when r = 50 nm (0.05 mm) and H = 5 nm, the force of attraction computes to $1.67 \cdot 10^{-12}$. Expressed in terms of the projection area πr^2 of the particle, this equates to a pressure of about 200 N m^{-2} and thus is negligible relative to the capillary forces. The gravitational forces (F_G) are also negligible, of course.

Of the resistance forces, only the deformation force (F_D) is relevant, because the electrostatic and steric repulsion forces (F_S) are in the same order of magnitude as the van der Waals forces of attraction.

This leaves F_C and F_D as the only forces of relevance. For film formation, the following condition applies [67]:

$$F_C > F_D.$$

This condition can only be met if the particles are readily deformable, and that likely requires a temperature above T_g, i.e., the softened state.

Now, there is more to film formation than simple deformation of the particles. At this point, the film has no tensile strength. The particles need to fuse together. Two touching droplets coalesce once the first direct contact has been made with the droplet substance under the driving force of surface or interfacial tension. However, as explained above, experts disagree about the existence or amount of interfacial tension. The other cause of particle fusion – and probably the main one – is interdiffusion.

Interdiffusion proceeds rapidly only above T_g. For sufficient film strength, the molecules must become entangled across the original particle boundaries. This requires that the diffusion lengths be at least in the same order of magnitude of the entanglement lengths of the polymer molecules (see Chapter 5.1.6), the latter are much smaller than the particle diameters [22, 70]. Depending on the position of T_g relative to the service temperature, interdiffusion continues for days and weeks and in fact may never attain equilibrium [22, 66].

Interestingly, individual particles are discernible to a certain degree even after film formation is complete [64]. Nonvolatile adjuvants, such as emulsifiers, protective colloids, ionic initiator residues and possibly buffer salts accumulate mainly in the former gusset regions of the particles and have a detrimental effect on the film properties, especially water resistance.

6.2.1.3 Minimum film-forming temperature and coalescing agents

As previously explained, deformation and coalescence of the particles, along with interdiffusion, require that the particles in the film formation be in the softened state. This is the case when the film forming temperature T is above T_g. In practice, however, it is not the T_g of the dispersion or the polymer which is determined, but rather the lowest temperature above which the dispersion forms a coherent film. Details can be found in DIN 53787; the standardized measuring instrument is called an **MFFT bench**. MFFT is the abbreviation for minimum film-forming temperature. Below the MFFT, a cracked film or even (below the **white point**) a whitish powdery layer will form [10]. The MFFT is usually somewhat lower than the T_g of the pure polymer, because the dispersion water has a plasticising effect which varies with the hydrophilicity (polarity) of the polymer. For example, a polyvinyl acetate dispersion whose pure polymer has a T_g of 28 °C can have an MFFT of 16 °C [10]. Furthermore, the hydrophilic auxiliaries (emulsifiers, protective colloids) co-determine the MFFT, too, by virtue of their water-binding capacity – the higher the content of these in the dispersion, the lower is the MFFT [22].

Since many dispersion coating materials, such as masonry paints and wood stains for outdoor use, must lend themselves to application at temperatures slightly above 0 °C, a major problem arises: on one hand, the MFFT must be close to 0 °C, while, on the other, the coating must not become tacky at temperatures up to about 50 °C, as otherwise **blocking** (bonding between dry films one on top of the other) and/or irreversible soiling by adhering dirt particles may occur. Solving Equation 5.7 for $T - T_g$ and substituting 10^7 Pa s for η_T, which is the minimum viscosity for the blocking resistance [22], leads to the condition $T_g \geq T - 21$ K, i.e., the T_g in this example must be at least 50 - 21 = 29 °C. This contradiction is usually resolved - except in the case of interior paints - by adding at most 10 % of an auxiliary solvent (**film-forming agent**, **coalescing agent**), which softens the dispersion particles during film formation, but rapidly evaporates thereafter, thereby allowing the T_g to rise again.

The effect of a film-forming agent can be readily explained by the **three-phase model**. This states that the film-forming agent (FFA) distributes itself as follows:

Water phase $\leftrightarrow$ Particle exterior $\leftrightarrow$ Particle interior

The greatest MFFT reduction occurs when the FFA accumulates mostly in the exterior areas of the particles, since that is where coalescence takes place. It follows that the solvent properties of the FFA must be matched to the type of polymer in the dispersion. Typical FFAs are esters such as 2,2,4-trimethyl-1,3-pentanediol monoisobutyrate ("Texanol"), ether alcohols, such as (di)ethylene glycol monobutyl ether (butyl (di)glycol), or glycols such as ethanediol or propanediol (ethylene or propylene glycol) [71].

Apart from the plasticising effect of an FFA, film formation depends heavily on the combination of water and polymer. The FFA should evaporate more slowly than the water and thus accumulate in the dispersion during film formation, yet escape quickly from the film after coalescence in order that the final hardness may be obtained rapidly. This is where air humidity is important. At low humidity, the water evaporates quickly and the FFA tends to accumulate. At high humidity, the converse happens. The FFA must be matched to the application conditions - which can be viewed as application windows (see Figure 2.13) - in addition to the dispersion itself [72].

6.2.1.4 Other ways of lowering the MFFT

It might be assumed that the particle size should have an impact on the MFFT. However, if anything, only a weak reduction in the MFFT with falling particle size has been observed; a particle size reduced by a factor of eight reduces the MFFT by five to ten degrees [65]. In some cases, no effects, and indeed even the opposite effects, have been observed [22, 67]. A targeted and significant reduction in the MFFT is therefore not achievable via changes in the particle size.

Aside from the very rare co-use of true external plasticisers, such as practically nonvolatile phthalates, phosphates, etc., there remains only the possibility of making or blending dispersions with particles of different T_g or, preferably, building up dispersion particles by corresponding polymerisation methods in a deliberately non-uniform manner. This enables particles to be generated that become steadily harder from outside to inside, i.e., have a hardness gradient, or truly heterogeneous (multiphase) structures. The latter have names that reflect their basic structure: acorn, crescent, occlusion, strawberry or core-shell morphology [14].

The most common structure is the **core-shell** morphology with a soft shell and hard core. The soft shell of low T_g provides the coalescence, while the hard core of high T_g contributes the film hardness. One example is that of acrylate dispersions, which have an MFFT of around 0 °C and also good blocking resistance and high gloss at up to 80 °C. The high blocking resistance is ascribed to the shielding of the soft phase on the surface by the bonded hard phase cores [73].

It should not go unmentioned that it is possible to increase the T_g and thus the resistance of emulsion paint films during and after physical film formation by additional interparticulate crosslinking (see Chapter 8.3, reactions 14 to 18).

6.2.1.5 Pigmented dispersion coating materials

In pigmented emulsion paints or coatings, the film-forming processes are in principle similar to those for the pure polymer film. Any MFFT quoted refers only to the pure dispersion. Where pigmentation occurs the MFFT increases as a function of the formulation [66].

A fundamental distinction must be made between the sub-critical pigment dispersion coatings or paints, and the super-critical emulsion paints. The former usually have PVC of less than 25 % [14]. The properties are mainly determined by the binder, and the same relationships between MFFT, blocking resistance and other film properties apply as in the pure dispersion films [73, 74].

In super-critical interior emulsion paints that have PVC values of 60 to 90 %, the dispersion particles encapsulate the markedly or substantially larger pigment and filler particles, bonding them to each other during film formation. The binder has an indirect influence only on the properties. In the past, it was believed that a dispersion with an MFFT of less than 5 °C would lead to insufficient pigment and filler binding – measurable as poor **wet scrub resistance** or **washing** and **scrubbing resistance** – and would lead to cracking of the emulsion paint, and that a dispersion with an MFFT higher than 5 °C required a film-forming agent [66]. More recent types of dispersion, such as terpolymers of vinyl acetate, acrylate and vinyl versatate ("VeoVa") [75] or vinyl acetate-ethylene copolymers [76], in combination with other formulation optimisations, facilitate the production of **emission- and solvent-free emulsion paints** that rival the performance of the classic paints.

6.2.2 Waterborne secondary dispersions

Most secondary dispersions, such as epoxy, polyester, silicone and alkyd resin dispersions or emulsions are relatively low molecular and must therefore be crosslinked. During the preceding water evaporation, the colloidal particles must rapidly coalesce into a film. For, if crosslinking and thus solidification of the polymer occurs beforehand – i.e., at the particulate stage – a resilient film of sufficient tensile strength cannot form as the particles lie beside each other largely unattached. This is particularly the case for 2-pack waterborne systems (see Chapter 8.4.1.2 and 8.4.2.3).

An example of a secondary dispersion that often undergoes purely physical film formation is that of **polyurethane dispersions** (PUR or PU). The polymer particles consist of largely linear polyurethane-polyurea molecules of average molecular weights (M_n) of about 3 to 11 $\cdot 10^4$ g mol^{-1}, and occasionally much higher [77]; the particle size is usually below 0.2 μm. For stabilisation in the aqueous phase, they contain a sufficient quantity of pendant acid groups, generally carboxy groups, which form very hydrophilic ionic centres by neutralisation. Cationically and non-ionically stabilised PUR dispersions also exist [2, 19].

Film formation by a secondary dispersion differs from that by primary dispersions in two major respects:

- T_g of the dry film for soft grades lies between -70 and -40 °C [177]
- dispersion particles are extensively swollen with water and so are easily deformable
- PUR dispersions are self-emulsifying, i.e. thermodynamically stable

The low T_g leads to an MFFT of around 0 °C, whose lower level in the absence of solvent is limited only by the freezing point. The question immediately arises as to how such a soft film can be used at all when primary dispersions with much higher T_g values will block. The answer lies in the unusual polymer structure: hard **urethane** and perhaps also **urea segments**, which are packed together intermolecularly via hydrogen bonds to yield crystal-like structures and inhibit interdiffusion, and which are connected to each other via longer soft segments [78].

Hard, solvent-free PUR dispersions have recently become available which, despite a T_g of 50 to 60 °C, form films at 0 to 5 °C and do not block until the temperature reaches 160 to 180 °C [173]. The fundamental relationships governing purely amorphous polymers are very hard to discern here.

Thermodynamic stabilisation of the dispersions, which should counteract film formation, never materialises during drying for two reasons: first, the film dries out, and this leads to the collapse of the mutual electrostatic repulsion of the particles and, second, the neutralising agent may volatilise, thus causing the polymer molecules to lose their hydrophilicity.

Although films produced by physically dried PUR dispersions have good optical and mechanical properties (resilience), surpassing those of films made from primary dispersions, they are crosslinked using 2-pack technology or by stoving in order that the highest demands of the industrial and OEM sectors may be met. In addition to specific reactions for this, standard crosslinking chemical reactions chiefly find application here, such as oxidative drying, hydroxy + N-methylol (melamine resin, etc.), hydroxy + free or blocked polyisocyanate, epoxy + amine, and UV curing [19, 79, 80], as well as reactions 14 to 18 from Chapter 8.3, which are especially suitable for room-temperature curing of dispersions.

6.2.3 Emulsions

Emulsions, i.e. micro-heterogeneous distributions of a liquid binder in water, dry differently from high-molecular solid-in-liquid dispersions. As they are primary liquid substances, they need additional crosslinking or must crosslink spontaneously. As an example, consider the drying phase of an **alkyd resin emulsion** which is purely physical and which is followed by oxidative crosslinking (see Chapter 7).

Alkyd resin emulsions are colloidally stabilised with **external emulsifiers** (non-ionic and anionic surfactants) and partly also with amines/ammonia via salt formation. In some cases, a little organic solvent is added to lower the resin viscosity in the emulsion. They can be used for decorator's paints and industrial coatings, primarily for wood coating, as well as in corrosion protection [63].

Compared with acrylate dispersions, for example, primary alkyd emulsions (or the emulsified resins) have

- much lower average molecular weight (a few thousand g mol^{-1})
- lower viscosity
- lower Tg (-90 to -60 °C) [81]

Consequently, emulsions *inter alia* do not have an MFFT.

The first phase of water evaporation and the mutual approach of droplets proceed on the same lines as the drying of a dispersion. Incipient coalescence of droplets, which occurs much more easily here than in the case of high-molecular dispersion particles, is followed by **phase reversal** from an **oil-in-water (O/W)** to a **water-in-oil W/O) emulsion** via a metastable, optically clear, homogeneous intermediate state at the critical volume ratio. The residual water forms tiny, weakly light-refractive droplets in the surrounding resin, and is only able to reach the surface by means of migration and diffusion. At the same time, slow oxidative crosslinking commences [81].

Under optimised conditions, the various phases can be seen temporarily by allowing droplets of an alkyd resin emulsion or an emulsion clear coat to surface dry on a glass plate (see Figure 6.10).

6.2.4 Non-aqueous disperse systems

Non-aqueous disperse systems include

Non-aqueous dispersions (NADs)

(generally polyacrylate) in an aliphatic hydrocarbon by way of the non-dissolving dispersant; high nonvolatiles content (80 %) is possible; for high quality, high solids topcoats; purely physical drying or thermal curing, are also combined with dissolved film formers.

Plastisols

Pasty dispersions of PVC (copolymer) powder – together with pigments and stabilisers – in a plasticiser (phthalate, adipate, phosphate, etc.) which is inactive at room temperature; some solvent may be added for viscosity adjustment; thermo-physical film formation, e.g. at 200 °C for 30 to 60 s (coil coating) or 180 °C for 10 min to yield thick, resilient layers.

Organosols

Dispersions of PVDF (polyvinylidene fluoride) powder or other fluoropolymer – together with pigments and acrylic resin – in a hydrocarbon (white spirit and the like); thermophysical film formation at, e.g. 30 to 60 s at 240 to 260 °C (coil coating) to yield a thin, flexible and extremely weatherable film.

While the physical film formation of an NAD likely resembles that of an aqueous dispersion, a special mechanism is at work in an organosol and (to a greater extent) a plastisol. This is known as **gelling**: at room temperature, the plasticiser is merely a dispersant. At elevated temperature, however, the polymer particles melt and the plasticiser diffuses into the melt droplets, which then swell extensively. Mutual touching of the droplets leads to coalescence, with interdiffusion and levelling. Finally, the film solidifies on cooling.

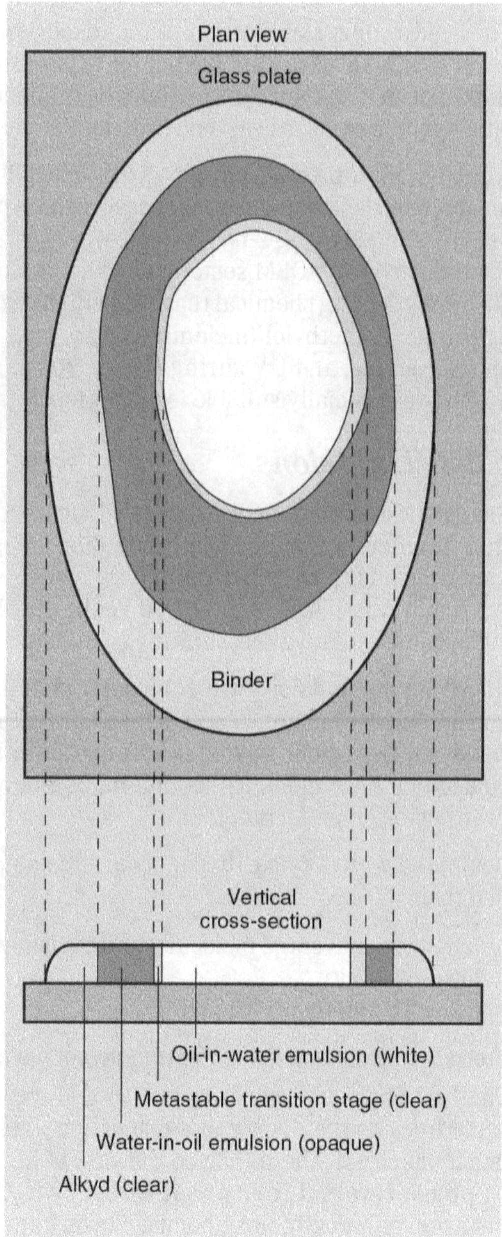

Figure 6.10: Colloidal surface drying stages in an alkyd resin emulsion droplet on a glass plate; from [81]

7 Oxidative crosslinking (oxidative drying)

7.1 General information and types of binders

The term oxidative crosslinking – traditionally called oxidative drying – refers to spontaneous curing of certain unsaturated film formers that crosslink in the presence of atmospheric oxygen.

Although it better describes the curing process than "oxidative drying", the latter will be used below because it reflects historical development and is the term used in the industry.

From ancient times, drying oils, notably linseed oil, have served as oxidatively drying film formers and, in siccative form (see Chapter 7.3), as clearcoats (**varnishes**). Most oxidatively drying binders nowadays, though, are made by making these oils react with other raw materials. The most important exemplar is **alkyd resins**, which are defined as polyesters that have been modified with oil or fatty acids (see below).

Oxidative drying has two significant advantages over many other crosslinking principles:

- coating materials are of the one-pack type, despite the fact that crosslinking takes place at room temperature
- crosslinker component (atmospheric oxygen) is freely available and does not need to be incorporated

Weaknesses and limitations are:

- crosslinking is relatively slow (takes hours)
- minor and downstream reactions increasingly cause yellowing and embrittlement
- unpleasant odours in the initial stages due to release of oxidative cleavage products
- little latitude as regards crosslink density and reaction conditions

The oils are without exception **triglycerides**, i.e., esters of a glycerol molecule and three fatty acid molecules. They are obtained predominantly from plant seeds and fruits, as well as aquatic animals and are refined to free them of unwanted materials (polysaccharides, proteins, lecithin, etc). Examples are:

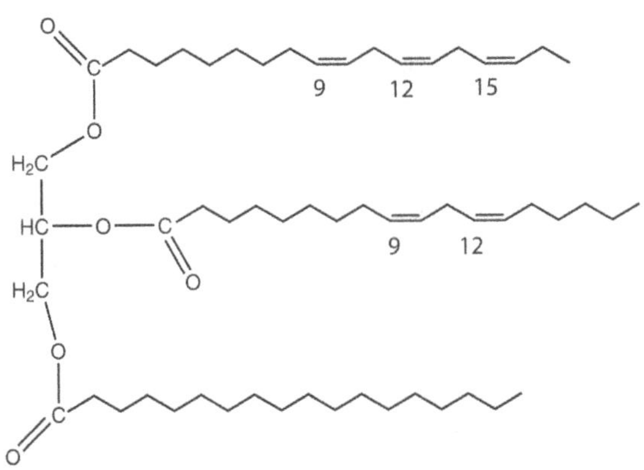

The numbers stand for the 9th, 12th and 15th carbon atoms, as numbered from the carboxy group (C_1)

The character of the oil is determined by the glycerols bound to the fatty acids. The fatty acids are generally long-chain monocarboxylic acids, usually containing 18 C atoms, that have up to three double bonds; fatty acids derived from fish oil can have up to five double bonds. The oils or fatty acids fall into one of the following classes as regards drying capacity.

In the fatty acid chain

- no double bond; oil is non-drying; example: lauric acid (found in coconut oil)
- one double bond; almost non-drying oil, example: oleic acid (found in peanut oil); reactive structure:

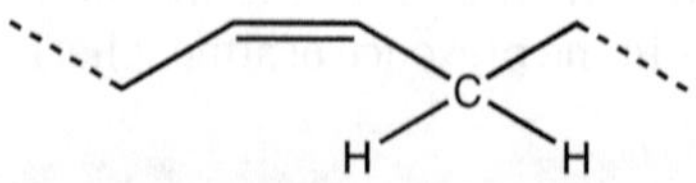

Monoallyl methylene group

- two **isolated double bonds**; oil is semidrying; example: 9,12-linoleic acid (found in soybean oil); reactive structure:

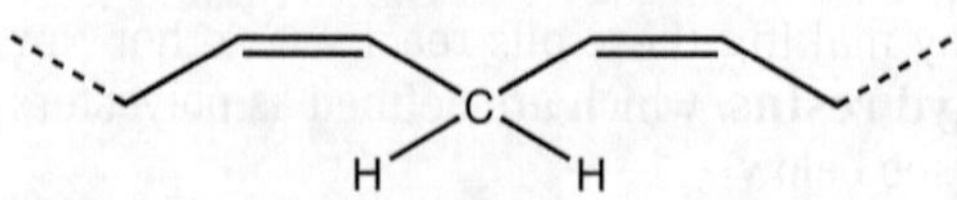

Diallyl methylene group

- three isolated double bonds; oil is drying, example: linoleic acid (found in linseed oil); reactive structure:

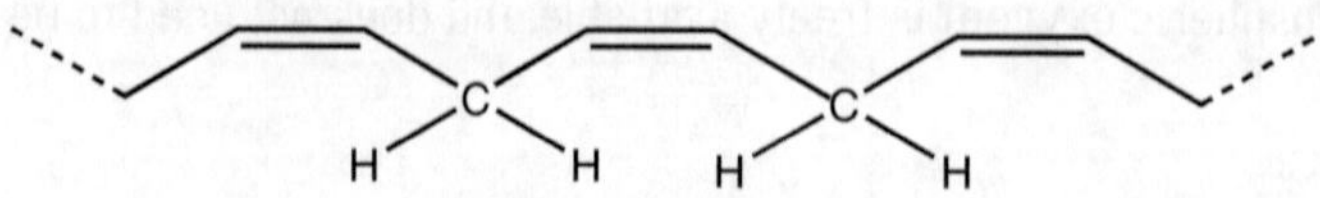

- two **conjugated double bonds**; oil is semidrying; example: 9,11-linoleic acid (found in castor oil); reactive structure:

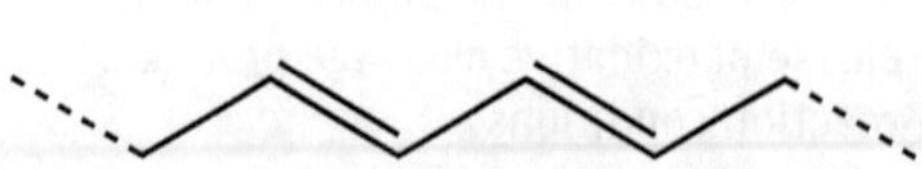

- three conjugated double bonds, oil is very fast drying; example: eleostearic acid (found in tung oil); reactive structure:

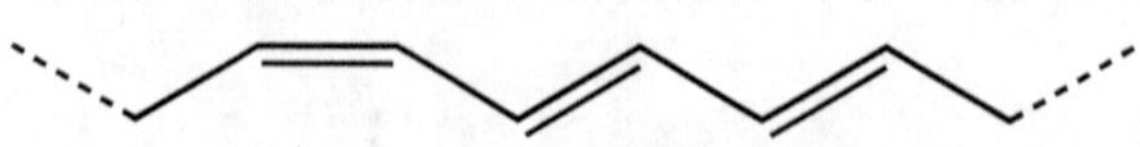

These distinctions can be condensed into two rules of thumb:

- the more double bonds there are in an oil molecule or the fatty acids which it contains, the faster and the more completely it dries
- conjugated double bonds "dry" much faster than the same number of isolated double bonds

A rough estimate of the content of double bonds or the degree of unsaturation is given by the **iodine value (IV)**. This is the mass of iodine in grams that can be chemically bound by 100 g of oil (or substance). Purely empirically, the following generally holds[18]:

IV above 150: Oil is **drying** or fast-drying (e.g. linseed oil, tung oil)
IV 100 to 150: Oil is **semidrying** (e.g. soybean oil, safflower oil)
IV below 100: Oil is **non-drying** (e.g. coconut oil, peanut oil)

[18] Literature limit; sometimes quoted as: IV 140

With **isolenic oils**, i.e. oils containing isolated double bonds, drying primarily starts at the neighbouring (di)allyl methylene groups, and not at the double bonds themselves (see Chapter 7.2).

Since oils are liquid 100 % film formers, they cannot undergo physical surface drying. Solidification can proceed only via oxidative crosslinking, which is relatively slow. Before alkyd resins became widespread, oils were often blended with physically drying binders - e.g. modified natural resins - and then **cooked** to achieve satisfactory compatibility. Nowadays, fatty acids are specifically incorporated into higher-molecular binders, such as polyesters, epoxy resins, phenol resins, polyacrylates and acrylic or polyurethane dispersions. The resultant solutions or emulsions/dispersions yield binders which undergo surface drying followed by oxidative curing. Moreover, oxidative drying can also be used as a downstream additional curing step for faster chemical crosslinking, e.g. by stoving.

Generally, as regards the crosslinking or curing of binders:

- The higher their initial molecular weights are and
- The more extensively branched their molecules are,
- The fewer new chemical bonds are needed for achieving high molecular weights and ultimately for crosslinking. Consequently, curing happens faster.

This explains why a drying oil with a molecular weight of just approx. 880 g mol^{-1}, just three fatty acid chains and just one branch requires a lot more new oxidative bonds for solidification or network formation than a long-oil alkyd resin which contains 65 wt.% fatty acid and consists of highly branched molecules with an average molecular weight of 5000 g mol^{-1} (see also Figures 5.6a and c). A rough calculation yields a figure of around eleven fatty acid chains per resin molecule. The resin molecule behaves roughly as if it were five oil molecules that are already oxidatively linked to each other. Such oxidized or **blown oils** are actually used as binders, too. Added to which, large molecules have an ability to undergo purely physical initial drying through solvent evaporation.

Oxidative drying capacity is essentially determined by the type and quantity of the fatty acids or other autoxidisable groups incorporated into a binder. Alkyd resins are therefore usually characterised on the basis of the type of oil or fatty acid and the oil content, expressed as **oil length**:

- less than 40 wt.% oil in the resin: **short-oil alkyd**
- 40 to 60 wt.% oil in the resin: **medium-oil alkyd**
- more than 60 wt.% oil in the resin: **long-oil alkyd**

The longer the oil (the greater the oil content), the lower is the contribution from fast physical drying and the greater is the extent of slow oxidative drying. Thus, fast-drying primers and undercoats are formulated on medium-oil alkyds and in some cases on short-oil alkyds, while more weatherable topcoats tend to contain long-oil alkyds.

The fatty acid chains are not the only carriers of oxidative drying capacity. Binders containing **allyloxy groups** ($-O-CH_2-CH=CH_2$), such as the unsaturated direct gloss polyesters (see Chapter 10.3.3) and relatively low-molecular cis-1,4-polybutadienes (polybutadiene oils) have it, too. In the allyloxy group, the methylene is just as susceptible as a diallyl methylene group to free-radical attack by oxygen. Polybutadiene chains contain numerous double bonds or allyl methylene groups:

cis-1,4-Polybutadiene (polybutadiene oil), $\bar{M}$ = 1500 to 3000 g mol^{-1}

Polybutadiene oils serve as a saponification-resistant substitute for fatty oils. They can be easily modified and converted, e.g., into a water-thinnable form. Oxidative crosslinking capability is also exhibited by butadiene copolymers, such as **styrene-butadiene dispersions**.

7.2 Mechanisms of oxidative drying

The chemical reaction mechanisms underlying oxidative drying have largely been elucidated, although certain details are not yet fully understood and are still the subject of research. The following mechanisms represent the consensus opinion in the literature and are provided as an aid to understanding film formation.

As the mechanisms vary to some extent, drying of binders containing isolated double needs to be treated separately from that of their conjugated counterparts.

7.2.1 Isolated double bonds

The simplified reaction scheme for the first part of drying is:

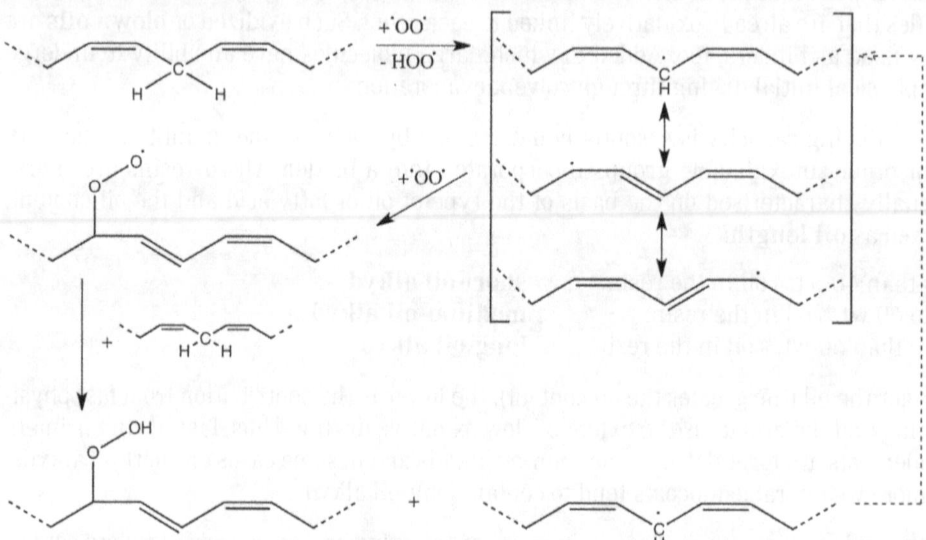

The mesomerism arrow ↔ connotes intermediate forms of a molecule which are formally created by delocalization of double-bond and possibly single electrons or free electron pairs. This delocalization is called mesomerism (or resonance) and leads to stabilisation (energy reduction) of the molecule, (see textbooks on organic chemistry for more details).

Autoxidation usually occurs at allyl or, more commonly, diallyl methylene groups. An oxygen molecule, which after all is a diradical (see Chapter 5.5.5), first abstracts a hydrogen atom from an activated methyl group, giving rise to a mesomerism-stabilised secondary radical. The radical produced by a diallyl methyl group is stabilised more than that produced by a monoallyl group and so it also reacts much more readily with oxygen. This is immediately followed by addition of oxygen to one of the mesomeric forms. The resultant **peroxy radical** in turn abstracts another hydrogen atom from a methylene group to form a **hydroperoxide** and another diallyl radical, to which oxygen again adds, etc. A **chain reaction** thus occurs, involving oxygen addition to yield hydroperoxides.

These descriptions are consistent with the following observations concerning the drying of linseed oil:

- first, during the **induction period** (incubation period), there is significant oxygen absorption (ca. 16 % mass fraction)
- there is virtually no increase in viscosity
- the double bond content does not decrease

At the beginning of this first reaction stage, any residual natural **antioxidants**, such as tocopherols (vitamin E), are degraded.

During the second drying stage comprising film formation, i.e., molecular enlargement and crosslinking, the hydroperoxides decompose into alkoxy and hydroxy radicals in a very slow process:

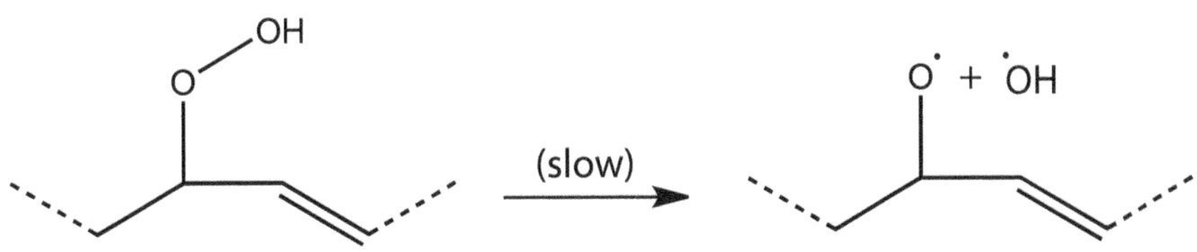

These radicals can in turn abstract hydrogen from fatty acid chains and so form secondary radicals (R˙) that can absorb oxygen

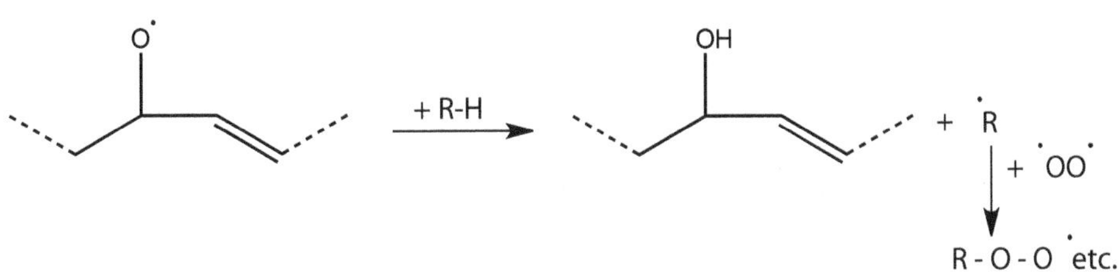

R-H is a further fatty acid chain
or they can combine with any other radicals present.

Overall, **radical combination** can occur in the following ways:

R-O•	+ •O-R	⟶	R-O-O-R	Peroxide linkage
R-O-O•	+ •R	⟶	R-O-O-R	Peroxide linkage
R-O-O•	+ •O-O-R	⟶	R-O-O-R + O_2	Peroxide linkage
R-O•	+ •R	⟶	R-O-R	Ether linkage
R•	+ •R	⟶	R-R	C-C linkage

Furthermore, the radicals add to double bonds, leading to further intermolecular linkages and loss of double bonds, e.g.

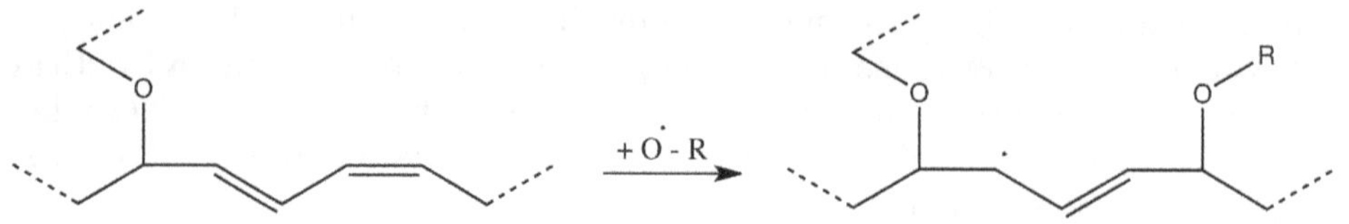

While ether linkages predominate in films dried at ambient temperatures (in air), the C-C linkages start to increase in number with increase in drying temperature, until they predominate in stoved films. Peroxide bridges are also formed in small numbers and are relatively stable in the polymer network. When linolenic acid ethyl ester was dried in the presence of siccatives (see Chapter 7.3), epoxy groups were also detected, the quantity of which peaked after about five days and was virtually zero after about 100 days. It is assumed that the epoxy groups reacted with carboxy groups [22].

The oxygen content of the film is a maximum at the end of the induction period and declines markedly with incipient solidification – in the case of linseed oil, it drops to a residual content of 10 to 12 % of the oil mass. This can be explained by cleavage of oxygen when two peroxy radicals meet.

The persistent unpleasant odour which occurs at the start of oxidative drying of isolenic oils or derived binders stems from a minor crosslinking reaction in which the fatty acid chains are released by free-radical cleavage of, for example, a linoleic acid chain:

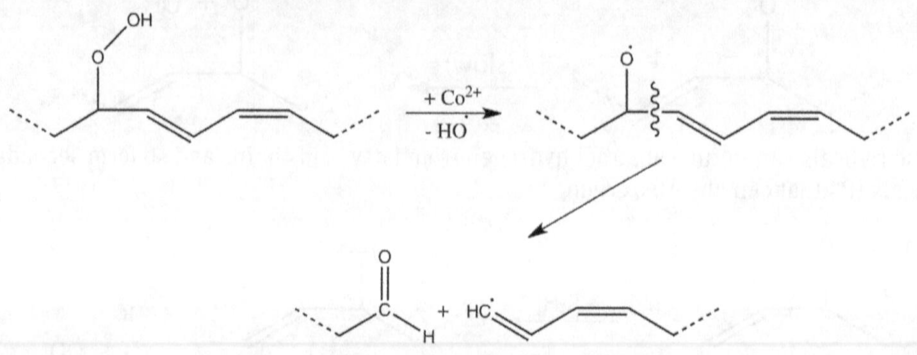

Various aldehydes, ketones and carboxylic acids are generally found as decomposition products [1].

7.2.2 Conjugated double bonds

The conjugated double bonds in so-called **conjuenic oils**, such as castor oil and tung oil, are directly attacked by oxygen or derived secondary radicals, leading to rapid oligomerisation by free-radical polymerisation (with frequent chain transfer), e.g.

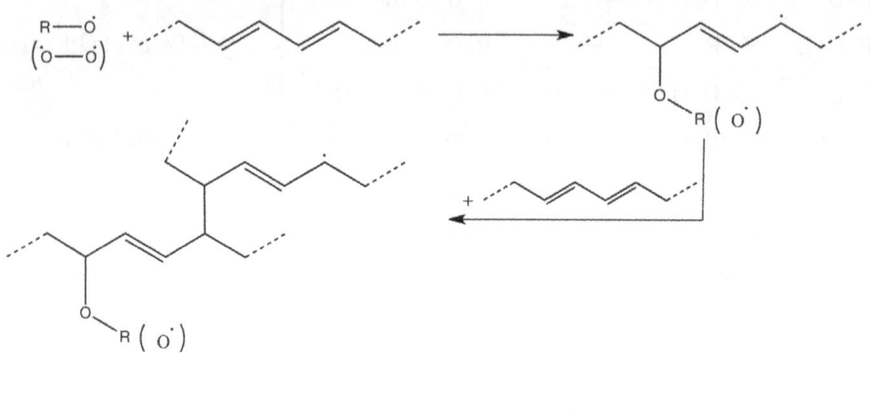

For the first addition step, a 1,4-addition of an oxygen molecule to yield a labile six-member ring is also assumed. In agreement with the formation of numerous C-C linkages, the films are observed to consume less oxygen during drying than do isolenic oil films.

Whatever the precise initial steps of the autoxidation process, what matters in practice is that conjuenic oil drying, as opposed to isolenic oil drying, is characterised by much faster film formation, superior weathering, water and alkali resistance, a lower tendency of the films to yellow and embrittle - and all as a result of the high content of C–C bonds. Figure 7.1 shows a schematic diagram of a network formed by a crosslinked conjuenic oil film.

7.3 Siccatives

A linseed oil film (isolenic oil) dries in approx. 125 hours at 25 °C. If the right amount of siccative (**drying agent**, dryer) is added beforehand, i.e., catalytic metal compounds, the same degree of drying can be attained in about $2^1/_4$ hours. A tung oil film (conjuenic oil) dries about twice as fast again. In both cases, therefore, oxidative drying is greatly accelerated by siccatives.

Siccatives (for non-waterborne) coating materials are salts of organic acids and various metals. Other metal compounds, such as certain complexes, are sometimes used. The nature of the organic acid primarily determines the siccative's solubility properties. The hitherto popular fatty acids and resin acids have now been widely superseded by naphthenic acids (petroleum-based alicyclic monocarboxylic acids) and especially the synthetic 2-ethylhexanoic acid (isooctanoic acid).

The metal [82] contained in a siccative determines whether it is a

- **surface drier:** cobalt (Co), manganese (Mn), vanadium (V), iron (Fe) and cerium (Ce)
- **"through-drier"**(coordination drier) zirconium (Zr), aluminium (Al), bismuth (Bi), barium (Ba), etc.
- **auxiliary drier** (promoter): calcium (Ca), potassium (K), lithium (Li), zinc (Zn) (inhibitor)

Another classification distinguishes between **primary driers**, which are identical with the surface driers above, and secondary driers, which comprise all other metal salts which are active in any way.

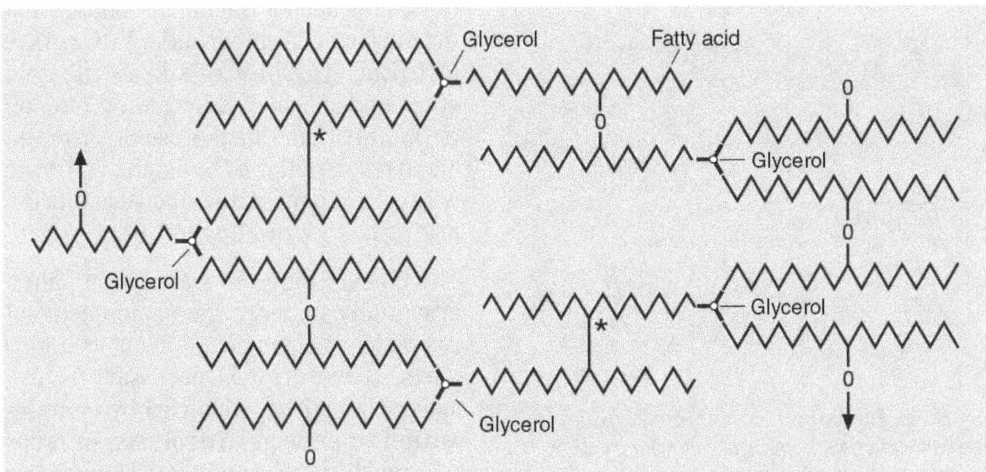

Figure 7.1: Crosslinked conjuenic oil (schematic), C-C linkages marked with an asterisk; from [6]

The metals of the surface driers are capable of alternating between two (or more) oxidation states, which is how their catalytic effect arises. For the most active metal, Co, the redox equations are:

$$R\text{-}O\text{-}O\text{-}H + Co^{2+} \longrightarrow R\text{-}O\cdot + OH^- + Co^{3+}$$
$$R\text{-}O\text{-}O\text{-}H + Co^{3+} \longrightarrow R\text{-}O\text{-}O\cdot + H^+ + Co^{2+}$$

$$2R\text{-}O\text{-}O\text{-}H \longrightarrow R\text{-}O\cdot + R\text{-}O\text{-}O\cdot + H_2O$$

It takes a long time for the hydroperoxides formed during the first drying stage to decompose into radicals that can link molecules together, because the activation energy for forming two radicals is high. If, instead, only one radical and the low-energy hydroxide ion are formed, the activation energy is much lower and decomposition proceeds commensurately faster (see Figure 7.2). An **oxygen-transfer mechanism** has also been postulated for primary driers in addition to this **redox mechanism** [83].

The use of Co has come in for criticism for toxicological reasons and it is being replaced in some cases by Mn and V, which not only are less active but also cause stronger discoloration [84]. Mn has generally been used in combination with Co up to now, with the Co providing very rapid **surface drying** while the manganese contributes better **through-drying**.

The through-driers and auxiliary driers (secondary driers) accelerate the effect of the **surface driers** (primary driers) and make for more uniform curing across the whole layer. The effects stem either more from ionic crosslinking (salt crosslinking) of the binder carboxy and hydroxy groups or from an as-yet not fully understood oxygen-transport mechanism acting in coordination with the surface driers. The most effective through-drier, lead, is being replaced by Zr for toxicological reasons [14].

The **single-metal driers** are used in quantities of 0.03 to 0.2 wt.% metal (Al: up to 1 %), expressed in terms of the binder. The specific combination and dosing of the metals must be matched to the binder and the drying conditions; ready-to-use metal combinations (**combination driers**), e.g. of Co, Mn and Zr, are available for standard formulations.

Excessively fast surface drying harbours the risk of **wrinkling**. The uppermost film layer shrinks as it crosslinks and thus prevents oxygen transport inside the film. Subsequent (slow) through-drying of the film – perhaps boosted by solvent evaporation – causes the deeper fraction to contract and the surface skin to wrinkle. In **wrinkle (crinkle) finishes**, this is a deliberate effect and is achieved by adding tung oil to the finish and/or the use of castor oil alkyd resins, followed by extensive drying with cobalt. Temperature increases during drying boost wrinkling [60, 85].

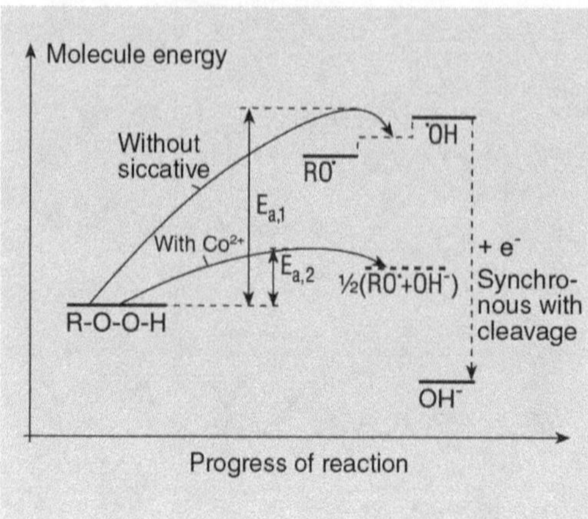

Figure 7.2: Energy profiles for hydroperoxide decomposition, with and without primary drier. Note the marked difference in activation energies and the resultant differences in cleavage rates

The binders found in waterborne paints essentially undergo the same chemical reactions as their non-aqueous counterparts. However, it is necessary to use special siccatives, stabilised by **complexation** to prevent **hydrolysis**, in order that rapid loss of activity after incorporation into the coating material may be avoided. Additionally, somewhat higher doses are needed than in solventborne systems, as the radicals are less reactive in aqueous milieu. Typical formulation ingredients for waterborne systems, such as ammonia, amines, phosphates and anionic surfactants, can also impair the activity of the metals, especially that of cobalt [82].

A general rule of using siccatives is:

- keep the addition of siccative as low as possible because the redox-active metals promote continuous further oxidation, leading to embrittlement of the binder and hence of the paint.

7.4 Preventing skinning

In opened or partly full containers, the trapped air generally creates an oxidatively dried surface layer or skin on the oxidatively drying coating material. This problem is exacerbated by the fact that the siccatives prefer to accumulate on the surface due to their surfactant (soap-like) molecular structure [14].

Skinning is countered by adding **antiskinning agents**. Two different classes of substance are employed, each having a different mode of action: **antioxidants** and **complexing agents.** They are frequently used in combination with each other.

The antioxidants are **substituted phenols**, which act as reducing agents by scavenging oxygen-containing free-radicals. They have low volatility and evaporate at the latest during stoving. In air drying, they act as temporary **inhibitors** by prolonging the induction period. This can have a beneficial effect on the end properties of the film [14].

The complexing agents are almost exclusively **oximes**, which are condensation products formed between ketones or aldehydes and hydroxylamine. By far the most common representative is methyl ethyl ketoxime (butanone oxime). The oximes are about as volatile as solvents. This is beneficial in two ways. First, the oxime is also present in the vapour space of the container, with the result that it can readily act on the paint surface and, second, it evaporates after application, thereby allowing oxidative drying to set in rapidly.

The conventional explanation of how oximes work is as follows. After oxidation from Co(II) to Co(III), the cobalt must lend itself readily to reduction again. The oxime stabilises the trivalent state to the extent that it fails to revert to the divalent state in the conditions obtaining. Co(III) (Co^{3+}) has the electron configuration $3d^6$, and needs twelve electrons before it can attain the desired noble gas configuration $3d^{10}\,4s^2p^6$; these are supplied by three bidentate oxime ligands. Co(II) would be unable to attain the noble gas configuration. However, some doubt surrounds the redox mechanism in this simplified form - for further information, see [83].

7.5 Influences on oxidative drying

The rate of oxidative drying depends on all conceivable external factors, such as temperature, air humidity, ventilation, illumination, substrate and layer thickness. It is therefore not possible to state a precise, universal drying time for oxidatively drying coating materials. Added to which, formulation ingredients, such as siccatives, antiskinning agents and pigments also exert an influence. What follows is a listing of general facts only, no details are provided of any specific measuring results or theoretical background.

Alkyd resin films or coatings containing added siccative dry faster when

- temperature is high
- humidity is low
- layer thickness is low
- level of pigmentation is high (in the case of thick layers)

Furthermore, a heavy dependence on the substrate material has been observed. Drying proceeds much faster on steel, for example, than on glass [86].

Pigments can reduce drying ability by adsorbing and thereby deactivating siccatives when paints are in storage. Generally, the chemical properties of the pigment surface, e.g. the acid-base nature and the quantity of adsorbed water and other cleavable foreign substance, can have a complex influence on oxidative drying.

Rapid surface drying due to strong ventilation and/or high levels of or the wrong choice of siccative can retard through-drying. The reason is that the **barrier effect** of the surface skin prevents release of solvent from the inside and absorption of oxygen into the deeper layers.

Particular problems can arise in the coating of wood, especially of tropical origin. Exuding phenolic and other reducing **wood ingredients** are essentially antioxidants and thus can substantially retard oxidative drying.

8 Curing of liquid coatings by step-growth reactions

8.1 General

The term curing as applied to coating materials refers to crosslinking of the binder molecules. In short, curing is the same as crosslinking. In coating materials, the binder molecules are initially present as oligomeric or polymeric individuals. If they are high-molecular, they can form stable films without crosslinking, i.e., by physical drying.

The following disadvantages of physically drying coating materials or the resultant coatings can be overcome by targeted crosslinking:

- limited resistance to heat and organic liquids (solvents, etc.)
- mechanical property profile that is not always optimal
- high solvent content (VOC value) of truly dissolved films

The average molecular weights of the crosslinkable film formers lie in the oligomer range or slightly above it, i.e., below 10,000 g mol^{-1}. Thus, film formers have such low viscosities that they are more or less high solids solutions (nonvolatiles content > 50 %) or may even be processed as 100 % liquid systems. In water, low-molecular film formers take the form of emulsions or solutions – depending on the degree of hydrophilicity or miscibility with water, perhaps following neutralisation. High-molecular film formers form dispersions with water or other non-solvents, undergoing physical drying to films that can possibly be additionally crosslinked.

The crosslinking reactions are polyreactions, i.e., polyaddition, polycondensation or free-radical polymerisation (in the classical sense). Curing by free-radical polymerisation is discussed in Chapter 10. This chapter will look at only the first two types of reaction.

8.2 Polyaddition and polycondensation

Collision by molecules bearing at least two functional groups each in a mixture containing groups capable of reacting with each other in pairs without release of substances such as water, alcohol, ammonia or hydrogen chloride leads to molecular enlargement by **polyaddition**. One example is the formation of a linear polyurethane by reaction between isocyanate -(OCN)- and hydroxy -(HO-) to form a urethane-(O-C (O)-NH-) group[19]:

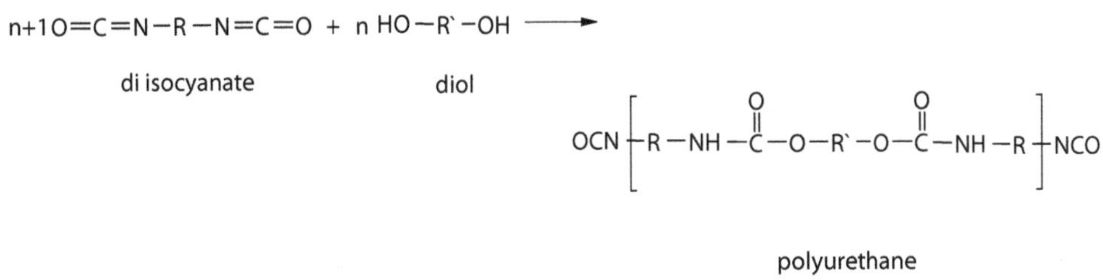

[19] It is common in linear structural formulae to write C(O) for oxygen atoms joined to the C atom by double bonds in order that they may be distinguished from C-O single bonds.

Peter Mischke: Film Formation in Modern Paint Systems

ISBN 978-3-86630-861-9

The hydroxyl groups add to the isocyanate groups at room temperature in an exothermic reaction. Branching occurs if the isocyanate and/or alcohol is higher than bifunctional and crosslinking happens if the polyaddition causes a certain average molecular weight to be exceeded, too.

If the polyreaction aksi gives rise to simple molecules such as those below, the reaction is a **polycondensation**. One example is the reaction between a dimethylol compound and a diol, with cleavage of water:

$$\underset{\text{diol}}{n{+}1\,HO-R-OH} + \underset{\text{dimethylol}}{n\,HO-CH_2-X-CH_2-OH} \longrightarrow$$

$$\underset{\text{polycondensate}}{HO\!\left[R-O-CH_2-X-CH_2-O-R\right]_n\!OH} + 2nH_2O$$

(X = urea-, melamine-, phenol groupe)

Normal alcohol hydroxy groups react with each other only in extreme conditions. In the **methylol group** (HO-CH_2-), the OH is activated by the core element X, however. The methylol groups in paint crosslinker resins (**complementary resins)** are for the most part etherified, as a result of which it is mostly alcohol which is released during polycondensation, not water.

Appreciable polycondensation requires elevated temperatures, generally above 120 °C (stoving), because the activation energy is high. Strong **catalysis**, e.g. with acids (protons) can lower the reaction temperature, even as far as room temperature. This is known as **acid curing**. Naturally, polycondensation reactions also yield networks if the binder molecules

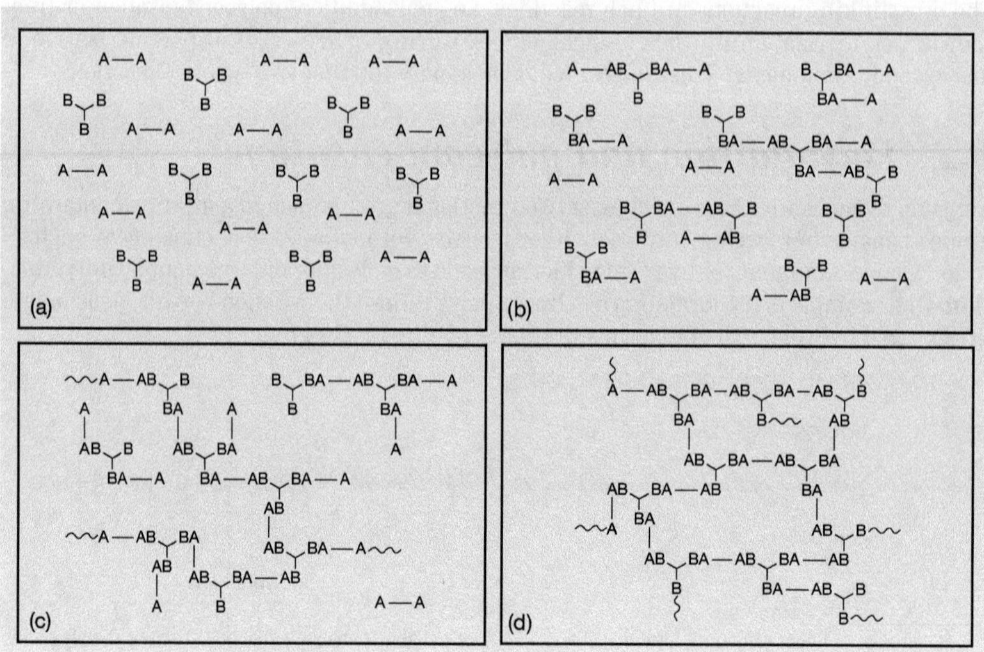

Figure 8.1: Formation of a network from bifunctional and trifunctional molecules bearing reactive groups A and B. a) prior to reaction; b) formation of branched chains; c) gel point exceeded; d) conversion/crosslinking complete

are more than bifunctional on average. Figure 8.1 presents a schematic view of network formation.

The following considerations apply whether or not the crosslinking is a polyaddition or a polycondensation. The important shared feature of both types of reaction, however, is that they are **step-growth polyreactions**. Whereas, in free-radical polymerisation, one molecule grows almost instantaneously in a **chain reaction** to yield a polymer molecule as soon as growth starts, the molecular weight of the molecules in a step-growth polyreaction increases incrementally - stepwise - over the entire period of the reaction. Each growth step is a discrete, fully self-contained, polar (ionic) reaction that follows the laws of standard organic chemistry (see Chapter 5.5).

8.2.1 Formal principles of molecular enlargement and crosslinking

Starting with the simplest scenario - bifunctional molecules (A-A and B-B) bearing reactive groups A and B in the ratio 1:1 - molecular enlargement can be formulated as follows:

A-A B-B A-A B-B A-A B-B A-A B-B A-A B-B

Conversion U = 0, mean degree of polymerisation $\bar{P}_n = 1$

↓

(e.g.) A-AB-B A-AB-BA-AB-B A-AB-BA-A B-B

U = 0.6 $$\bar{P}_n = \frac{2+4+3+1}{4} = \frac{10}{4} = 2.5$$

↓

A-AB-BA-AB-BA-AB-BA-AB-BA-AB-B

U = 0.9 $$\bar{P}_n = \frac{10}{1} = 10$$

where $\bar{P}_n$ is the arithmetic mean (number average) of the degree of polymerisation. The following general derivation

Equation 8.1

$$\bar{P}_n = \frac{1}{1-U}$$

shows that the degree of polymerisation, therefore, increases very slowly with increase in conversion, but then rises more and more sharply. The reason is that, at the start, the reaction mostly occurs between small molecules, whereas in the later stages ever-larger molecules are joining together to form much larger ones. The theoretical, extreme case of 100 % conversion corresponds to an infinite degree of polymerisation, which structurally and purely formally, would be achieved by ring closure in the last reaction stage.

For the general case that the number or molar ratio of the groups, known as the **stoichiometric ratio of reactants**

$$r = \frac{\text{Number of substiochiometric groups}}{\text{Number of substiochiometric}} \leq 1$$

is also less than one, the **Carothers equation** applies:

Equation 8.2:

$$\bar{P}_n = \frac{1+r}{1+r-2rU}$$

When r = 1, we obtain Equation 8.1 again. For U = 1 and, e.g., r = 0.9, we get

$$\bar{P}_n = \frac{1+r}{1-r} = \frac{1+0{,}9}{1-0{,}9} = 19$$

This means that lowering the stoichiometric ratio of reactants from 1 to 0.9 yields the value 19 for complete conversion rather than an infinite degree of polymerisation. This can be readily explained. The more unreacted groups that are present (are left over), the more chain ends there are, since every unreacted group forms a chain end. Two chain ends terminate one molecule each. Thus, as the number of free groups increases, there are increasingly more molecules present, and these are smaller on average, of course.

Reaction between exclusively bifunctional molecules does not bring about crosslinking. For crosslinking to happen there needs to be a certain content of trifunctional and higher-functional starting molecules present. Given a stoichiometric ratio of reactants of r = 1, which is also known as 1:1 **crosslinking**, we can easily derive the following:

The **mean functionality** of the mixture (paint) is

$$\bar{f} = \frac{\sum N_i f_i}{\sum N_i} = \frac{\sum N_i f_i}{N_0}$$

where N_i is the number (or number of moles) of molecules having functionality f_i and ΣN_i is the total number (N_0) of molecules. $\bar{f}\,N_0$ is the total number of functional groups present at time t = 0. For conversion U, $U\bar{f}\,N_0$ groups have reacted. As the crosslinking reaction proceeds, there is one molecule fewer in the mixture for every group pair (A+B) that has been converted and so $UN_{ex} = \frac{1}{2}\,U\bar{f}\,N_0$ molecules have disappeared by the time conversion has occurred. If N_t is the number of molecules still present at time t (conversion U), then

$$N_{ex} = \tfrac{1}{2}\,U\bar{f}\,N_0 = N_0 - N_t \qquad \Rightarrow \qquad N_t = N_0\,(1 - \tfrac{1}{2}\,U\bar{f}\,)$$

Since the mean degree of polymerisation is $\bar{P}_n = N_0/N_t$ (i.e. the mean number of monomer units or starting molecules per polymer molecule formed), this equation leads directly to

Equation 8.3:

$$\bar{P}_n = \frac{2}{2-U\bar{f}}$$

which is also known as the **modified Carothers equation** [21].

The degree of polymerisation then rises again with increase in conversion and becomes infinite at $U\bar{f} = 2$.

If, according to *Carothers*, attainment of the infinitely high degree of polymerisation corresponds to **gelation** or incipient through-curing of the mixture or paint film, the value of gelation conversion, called the **gel point** U_G (see also Chapter 5.1.7) is

Equation 8.4:

$$U_G = \frac{2}{\overline{f}}$$

For exclusively bifunctional starting molecules f = 2, and gelation would theoretically occur only at the practically unattainable conversion rate of 1 (100 %). The higher the mean functionality is above two, the lower is the gel point. Highly branched binder molecules bearing many reactive groups, such as long-oil alkyd resins or **dendrimers** (see Chapter 10.2.1.2) and **hyperbranched polyols** (see Chapter 5.1.7), require only a few new bonds between each other in order for gelation to occur. A mixture of triisocyanate and diol would have a mean functionality $\overline{f}$ of 1/5 · [(2 · 3) + (3 · 2)] = 2.4 and thus a gel point U_G of 0.83 (83 %). Until this conversion is attained, the paint film would still be capable of flowing, i.e. levelling and sagging.

An alternative gelation theory comes from *Flory and Stockmayer*; it is based on probability scenarios. The underlying idea is as follows: If one is looking for any of the numerous A (or B) groups in a reaction mixture – in this case, a paint film – the probability that this group has already reacted, i.e. that the molecule continues here, is equal to the conversion of the A (or B) groups. Through application of the addition and multiplication rules of probability calculations, the total probability of the occurrence of entire segment structures and thus also of branches can be calculated. The branching probability, in turn, is linked to the gel point by the assumption that, at the gel point, a (theoretical) non-terminating sequence of branching events occurs for the first time. This theory yields the following equation for the gel point U_G

$$U_G = \frac{1}{\sqrt{r(\overline{f}_A - 1)(\overline{f}_B - 1)}}$$

Equation 8.5::

$$\text{where } \overline{f}_A = \frac{\sum n_{A,i} f_{A,i}^2}{\sum n_{A,i} f_{A,i}} \quad \text{and} \quad \overline{f}_B = \frac{\sum n_{B,i} f_{B,i}^2}{\sum n_{B,i} f_{B,i}}$$

r	is the stoichiometric ratio of reactants
$\overline{f}_A$, $\overline{f}_B$	is the mean functionality of the A or B monomers
$n_{A,i}$, $n_{B,i}$	are the number of moles of A or B molecules of functionality $f_{A,i}$ or $f_{B,i}$

Note that the mean functionality here represents to a certain extent a weight average and not a number average. This reflects the fact that groups having higher-functional molecules have a greater probability of entering a reaction than those having molecules of lower functionality.

As a first example, let us calculate the aforementioned 1:1 crosslinking of a mixture of diol and triisocyanate: r = 1, $\overline{f}_A$ = 2, $\overline{f}_B$ = 3. This yields

Equation 8.5:

$$U_G = \frac{1}{\sqrt{1(2-1)(3-1)}} = \frac{1}{\sqrt{2}} = 0.707 \approx 71\,\%$$

This value is much lower than that returned by *Carothers* (83 %). The true gel point lies somewhere in between. The reasons are as follows [21]:

- *Carothers* assumes that gelation does not occur until $\overline{P}_n \rightarrow \infty$; however, it occurs earlier.
- *Flory* assumes that each branch is also a node that contributes to crosslinking. In fact, self-contained closed rings form from the branches; thus, conversion up until gelation must be higher.

As a second example, consider a – on average – trifunctional resin (containing A groups) that is cured with a crosslinker of following composition: 50 % of the molecules are trifunctional, 30 % are tetrafunctional, 15 % are pentafunctional and 5 % are hexafunctional. Let the stoichiometric ratio of reactants be 0.9:

$$r=0.9;\ \bar{f}_A=3,\ \bar{f}_B=\frac{0.5\cdot 3^2+0.3\cdot 4^2+0.15\cdot 5^2+0.05\cdot 6^2}{0.5\cdot 3+0.3\cdot 4+0.15\cdot 5+0.05\cdot 6}=3.96$$

$$U_G=\frac{1}{\sqrt{0.9(3-1)(3.96-1)}}=0.433\approx 43\ \%$$

After gelation, crosslinking continues with increase in extent of conversion:

- the average molecular weight of the network arcs $\bar{M}c$ decreases
- the thermoplastic, unbound molecular fraction (sol fraction w_s) decreases, while the fraction incorporated into the network (gel fraction w_g) increases
- the average molecular weight of the sol fraction $\bar{M}_{n,s}$ decreases, since larger sol molecules tend to be bound to the network more than smaller ones
- the mass fraction of free chain ends w_{End} rises initially after gelation, since the quantity of singly bonded sol molecules increases; thereafter, however, the free chain ends are increasingly bound.

This is illustrated in Figure 8.2.

To summarise:

- the higher the functionality of the binder molecules and the closer the stoichiometric ratio of reactants (r) is to 1, the
  - lower is the gel point, and the
  - higher is the crosslink density

The ultimate crosslink density, i.e. the network mesh width, depends not only on the functionalities and the stoichiometric ratio of reactants, since, e.g., three groups (f_i = 3) can be present on a very short or a very long molecule. This has (theoretically) no influence on the gel point, but it does affect the network arc length. The primary relevant variable here would appear to be the **equivalent mass**, i.e. the molecular weight fraction accounted for by a functional group. Moreover, branches lead to a denser network, too, since every branch becomes an additional node sooner or later.

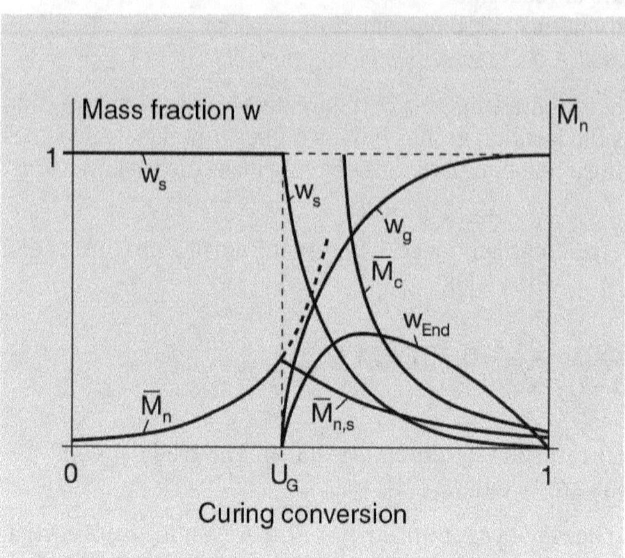

Figure 8.2: Molecular-structural changes in a paint film during crosslinking (schematic); see text for explanations; adapted from [12]

The stoichiometric ratio of reactants is directly related to the crosslinking ratio or the mole ratio $n_{hardener\ groups} : n_{resin\ groups}$ This is greater than one in the case of **over-crosslinking** but less than one for **under-crosslinking.** According to the rules above, given excess hardener, the crosslink density of ratio 1:1 would have to decrease again, since r falls below 1. As will be shown in the next chapter, however, curing conversion for 1:1 crosslinking does not generally reach 100 %. A small excess of crosslinker, therefore, can increase the curing conversion and thus the network density on account of the greater mobility of the relatively smaller crosslinker molecules. Furthermore, the crosslinker molecules may enter into minor or downstream reactions to form additional network bonds. This occurs with urethane systems in which excess isocyanate groups can react with atmospheric humidity or the paint film to yield **urea groups** and at elevated temperature and/or in the presence of catalysts with any urethane groups already formed to yield **allophanate groups** [7, 19, 22]. In epoxy-amine systems, a 10 to 20 % amine excess (more precisely: excess of NH groups) can benefit the curing rate and film properties [14].

8.2.2 Fundamental physicochemical principles of crosslinking

Once a crosslinkable, solventborne or waterborne coating material has been applied, it starts undergoing physical drying as described in Chapter 6. The concentration of the functional groups (A, B) in the film rises initially and the groups have almost unhindered access to each other thanks to thermal molecular movement and resultant diffusion in the as-yet liquid film. What happens next depends on the reactivity.

In a 2-pack system, the curing reaction begins as soon as the base paint and hardener are mixed, more precisely it occurs not just in the film, as desired, but unfortunately also in the mixture still waiting to be applied. Consequently, that mixture has a limited pot-life. In conventional 2-pack paints, the reaction in the mixture causes a steady rise in the viscosity, the measurement of which can serve as an indicator of the end of the potlife. This is frequently deemed to have occurred when the initial viscosity has doubled. In waterborne, 2-pack systems, however, there is generally no significant relationship between the start of the reaction in the mixture and the viscosity; in such cases, the pot-lives indicated by the paint manufacturer must be followed strictly (see Chapters 8.4.1.2 and 8.4.2.3).

1-Pack systems harden either only at greatly elevated temperature (stoving) or – where polymerisable – by irradiation (see Chapter 10.2). Oxidative drying (see Chapter 7) and moisture curing (see Chapter 8.4.1.3) are exceptional cases that make air-drying 1-pack systems possible.

A step-growth curing reaction is composed of a transport or diffusion stage and a reaction stage.

$$\sim\!\sim A + B \sim\!\sim \underset{k_{-D}}{\overset{k_D}{\rightleftarrows}} \sim\!\sim A\cdots B \sim\!\sim \xrightarrow{k_R} \sim\!\sim A-B \sim\!\sim$$

Groups A and B must collide in order to be able to react; this reaction is reversible. It is from this loose A···B compound that the actual, essentially irreversible, chemical reaction to afford the product A-B proceeds, i.e. a crosslinking node; the reaction comprises one or more elementary steps involving the relevant transition stages and activation energies.

From the overall rate constant $r_{tot} = k_R[A\cdots B]$ and the relationship $k_D[A][B] = k_{-D}[A\cdots B] + k_R[A\cdots B]$, and assuming a steady state, we get

Equation 8.6:

$$r_{tot} = \frac{k_R k_D}{k_{-D} + k_R}[A][B]$$

and by comparison with $r_{tot} = k_{tot}[A][B]$ (equation defining the overall rate constant):

Equation 8.7:

$$\frac{1}{k_{tot}}=\frac{k_{-D}}{k_D k_R}+\frac{1}{k_D}$$

If diffusion is very fast initially, then $k_D >> k_R$, and Equation 8.7 simplifies to

$$k_{tot}=\frac{k_D}{k_{-D}}k_R=K k_R$$

with the equilibrium constant $K=\frac{[A...B]}{[A][B]}$

A pre-diffusion equilibrium sets in quickly and the overall reaction rate constant k_{tot} is dependent virtually only on k_R. The latter, in turn, obeys Arrhenius' equation (Equation 5.21) for temperature dependence. This is called **reactivity control**.

As a result of solvent or water evaporation, molecular enlargement and incipient crosslinking, the T_g of the film increases continuously; the free volume sinks and the diffusion rate of the binder molecules or their segments or chain ends bearing reactive groups falls accordingly. In short: the A and B groups now collide each other more slowly and less often as the matrix solidifies more and more. Ultimately, $k_D << k_R$, i.e. the converse of the outset. In the denominator of Equation 8.6, the constant k_{-D}, which, like k_D is also very small relative to k_R, can be neglected. In that event, $k_{tot} = k_D$. The overall reaction is subject to **diffusion control**.

Equations 5.7 (WLF equation) and 5.18 can be combined to yield an equation describing the dependence of the diffusion constant on the temperature:

$$\ln\frac{D_{T_g}}{D_T}=\ln\frac{T_g}{T}-\frac{40(T-T_g)}{52+T-T_g}$$

This yields the theoretical information that the diffusion constant D_T at temperature T tends towards zero if the T_g approaches the temperature T + 52 K from below. T is the current film temperature, i.e. the curing temperature. In practice, the crosslinking rate begins to decline on account of diffusion once the T_g exceeds a temperature of T - 10 K [22]. If T_g moves into the range between 25 and 35 K above T, the diffusion rate and thus the reaction rate become too small to measure [12].

But even if T_g always remains markedly below the curing temperature, total conversion is not achieved for precise 1:1 crosslinking, because ultimately there are still functional groups which are immobilised in the network and cannot find a partner group; this is called the **steric** or **topological limit (U_{top}).** For rigid (hard) diepoxy diamine systems, this limit occurs at between 90 and 95 % for 1:1 crosslinking. If the stoichiometric ratio of reactants or crosslinking ratio deviates substantially from 1:1, an appreciable topological limit is never reached, however [12].

The T_g of a molecular network increases with increase in crosslink density and thus with increase in conversion of the functional groups (curing conversion). The rise in T_g due to crosslinking may be described empirically by [12]:

$$\frac{\Delta Tg}{K} = \frac{3.9 \cdot 10^4}{\frac{\bar{M}c}{gmol^{-1}}}$$

ΔT_g is the rise in the glass transition temperature due to crosslinking
$\bar{M}_c$ is the average molecular weight of the network arcs

This yields the following guide values:

- highly crosslinked system: $\bar{M}_c \sim 500\ gmol^{-1} \Rightarrow \Delta T_g \approx 80\ K$
  Example: 100 % 2-pack epoxy (liquid)
- Moderately crosslinked system: $\bar{M}_c \sim 2000\ g\ mol^{-1} \Rightarrow \Delta T_g \approx 20\ K$
  Example: polyester stoving paint

The T_g values of crosslinked paint films (when dry) are predominantly between 40 and 120 °C, a fact which compares well with the calculated rises in T_g.

From the various relationships that have been explained thus far, the following practical conclusions can be drawn: For high hardener conversion and thus the highest-possible crosslink density,

- the functional groups must be capable of reacting to completion within the curing time. The reactivity can be increased by raising the temperature and/or possibly adding catalyst
- the curing temperature must be substantially higher than the T_g of the fully cured film.

In oven drying, stoving time and stoving temperature have mutually opposing influences. For a given curing reaction, the relationship between stoving duration and object temperature is given by Arrhenius' Equation (5.21):

$$\ln\frac{t_2}{t_1} = \frac{E_a}{R}\left(\frac{1}{T_2} - \frac{1}{T_1}\right)$$

t_1, t_2 are the stoving times at temperatures T_1 and T_2 (in K)
E_a is the effective activation energy
R is the general gas constant (8.314 J mol^{-1} K^{-1})

A temperature increase of 10 K shortens the stoving time (holding time at object temperature) of most paint systems by 30 to 45 %, the exact figure depending on the starting temperature [6]. If the temperature and/or the stoving duration is too low and/or short, the reaction or crosslinking does not go to completion, with the result that the film is lacking in hardness and durability. This is called **understoving.** Conversely, **overstoving** leads to embrittlement and possibly yellowing of the film due to further crosslinking, side-reactions and perhaps chain cleavage. These relationships are shown in Figure 8.3.

Modern methods of physical analysis, such as **DMA (dynamic mechanical analysis)**, **DSC (differential scanning calorimetry)** and **DEA (dielectric analysis)** can be used to determine the optimum stoving conditions in advance [87-89]. Instrument makers also supply software for calculating the chemical kinetics of paint curing from the readings so that curing can be modelled in full.

The problem of diffusion-induced hindering of curing really only occurs with air-drying 2-pack systems. The T_g of the ready-to-use coating is generally much higher than room temperature. In an extreme case, the T_g is already so high prior to crosslinking – due to physical drying and molecular enlargement alone – that the paint film remains in the uncross-

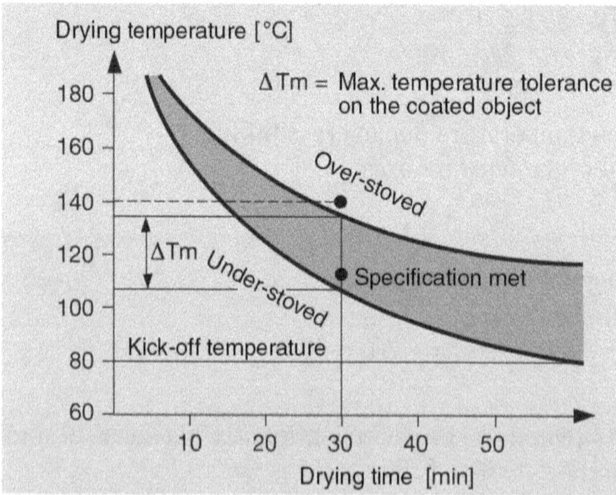

Figure 8.3: Typical scope for processing stoving paints in a dryer; from [7]

linked or slightly crosslinked state (**B state**) for a long time [12, 19]. Normally, the T_g does not rise above room temperature until extensive curing conversion has taken place. Possibly, residues of solvent lower the T_g, which would increase curing conversion, but would also harbour the danger of solvent retention. 2-Pack systems therefore continue to improve in durability over a protracted period (in the order of months) due to post-curing and perhaps solvent emission; this process can be accelerated by heating the film above the T_g of the dry film [22]. Figure 8.4 illustrates the relationship between rising T_g and curing conversion.

8.3 Important crosslinking reactions (selection)

A number of annotated, important crosslinking reactions in coating technology are presented below in the form of simplified reaction equations, inasfar as these do not involve oxidative crosslinking or free-radical polymerisation (see Chapters 7 and 10).

Notes on the chemical formulae:

- in each case, only the chemical reaction between the reacting functional groups is shown. For crosslinking to occur, the reacting molecules must be bifunctional or higher-functional; see Chapter 8.2.1
- the chemical backbones (polymer, oligomer, monomer) bearing the groups are not represented by "R" or wavy lines, bur rather symbolised by "Resin" or "Cross" (for crosslinker). "Polymer" is written wherever – as in the case of a polymer dispersion – the backbone need not be a resin (in the strict sense). "R" dennotes a small, non-functional group
- substances listed under the reaction arrow are typical catalysts (incorporated as additives); "Cat" stands for catalyst. "Cat" in parentheses signifies that the catalyst is not always needed
- "1P" stands for 1-pack and "2P" for 2-pack. 2-pack does not rule out additional curing at elevated temperature (forced drying or even stoving)
- A short arrow pointing upwards indicates escape of low-molecular reaction products

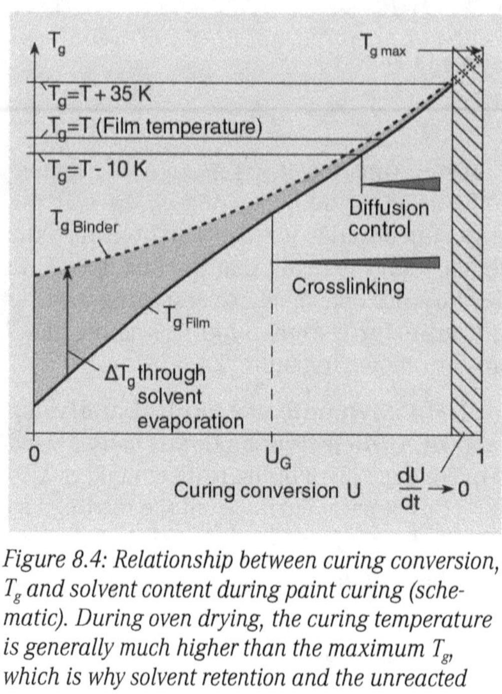

Figure 8.4: Relationship between curing conversion, T_g and solvent content during paint curing (schematic). During oven drying, the curing temperature is generally much higher than the maximum T_g, which is why solvent retention and the unreacted fraction both tend towards zero (U_G is the gel point)

1) Resin polyol (1) (short/medium-oil alkyd, polyester polyol, polyacrylopolyol, solid epoxy resin) + **formaldehyde condensation resin** (2) (melamine, urea, benzoguanamine, glycoluril, phenol resin):

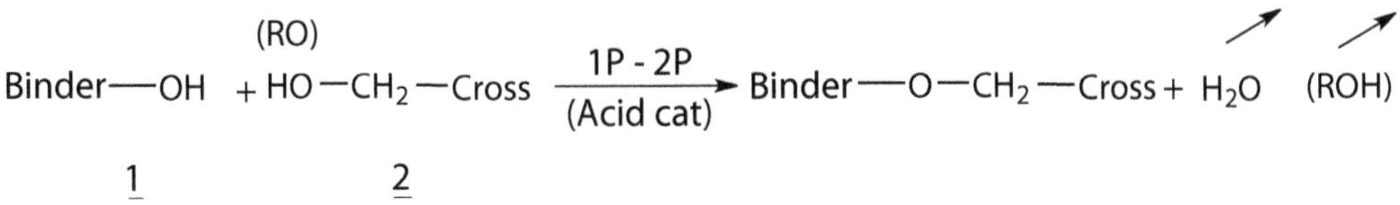

HO-CH2: Methylol group activated by N atom or benzene ring
(active N-methylol groups, perhaps in acrylic resins and acrylic dispersions, as well)

1′) Parallel self-crosslinking of varying extent:

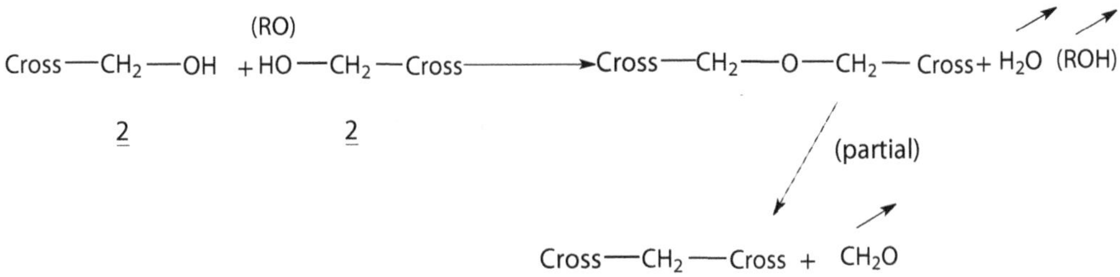

CH_2O: Formaldehyde

1″) Parallel, perhaps with carboxy groups (slower than with OH):

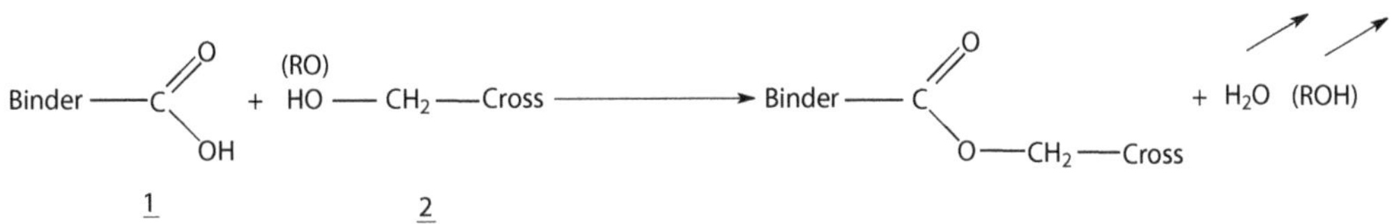

2) Resin polyol (1) + **polyisocyanate** (2):

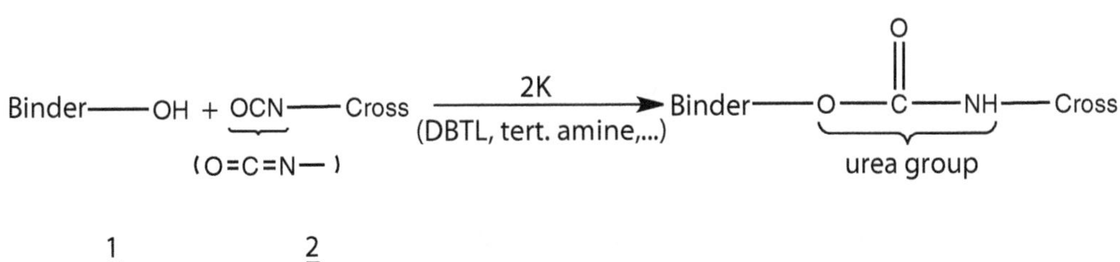

DBTL: Dibutyl tin dilaurate. (aromatic isocyanates: no catalyst)

3) Resin polyol (1) + **blocked polyisocyanate** (2):

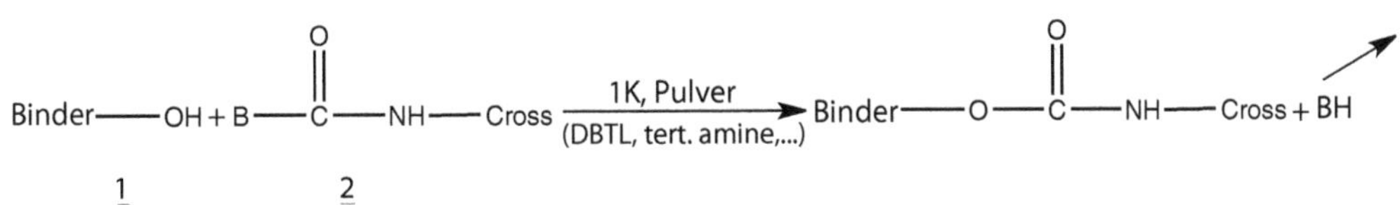

BH: Blocking agent, such as butanone oxime, 3,5-dimethylpyrazole, ε-caprolactam and others

4) Resin polyol (1) + **polyuretdione** (2) as non-cleaving, blocked polyisocyanate:

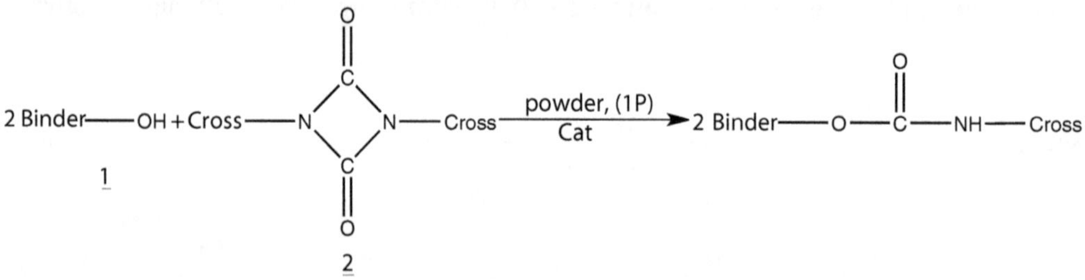

(primarily for powder coatings)

5) Isocyanate prepolymer (1) + **atmospheric humidity** (2):

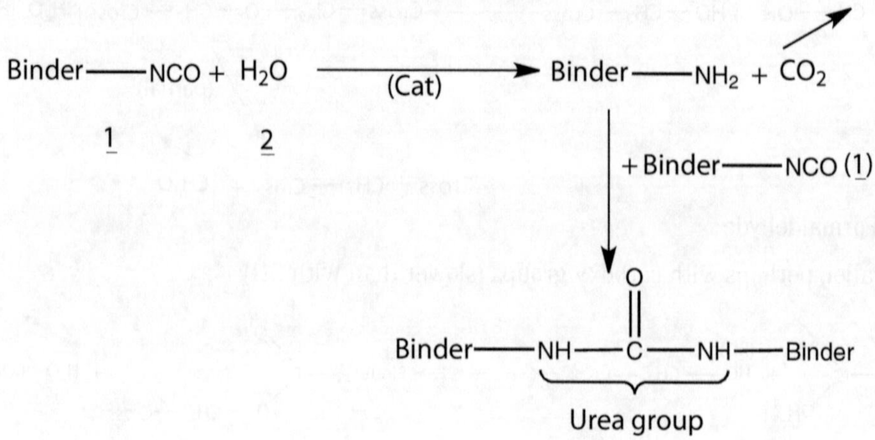

"Moisture curing". Risk of blistering at high layer thickness owing to cleavage of CO_2

6) Epoxy resin (1) + **amine crosslinker** (2) (amine, amine adduct, polyamidoamine etc):

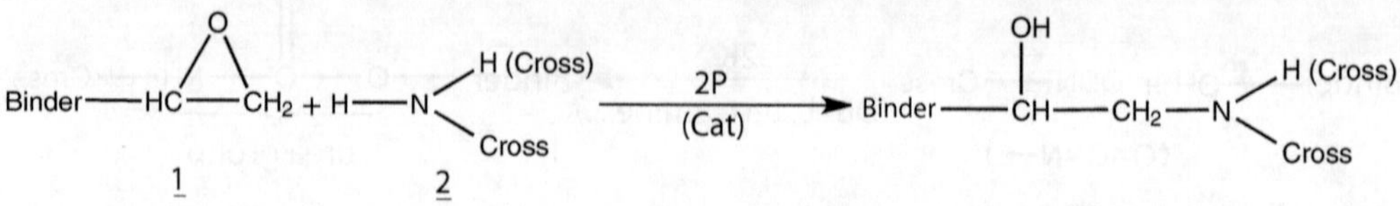

Parallel: anionic oligomerisation. Catalyst addition depends on reactivity of the crosslinker

7) Epoxy resin (3) + **polyketimine** (1) with atmospheric humidity (2):

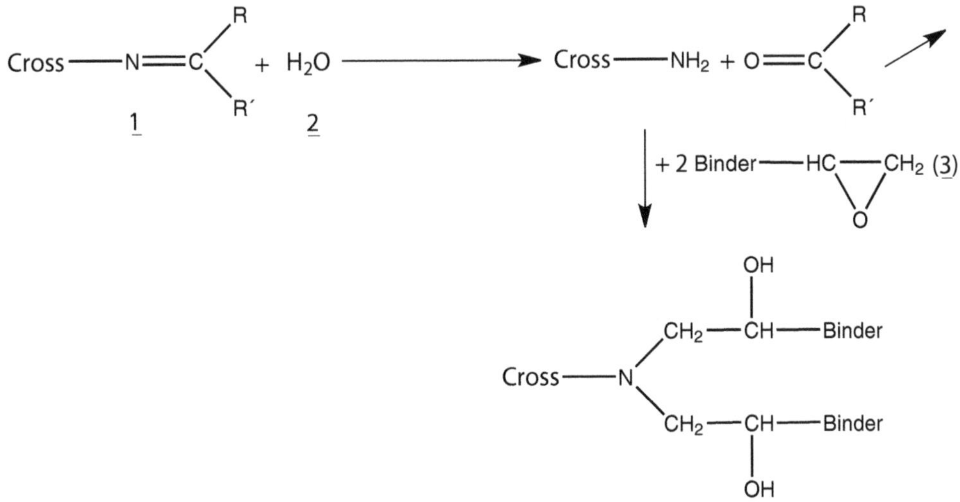

Potlife is longer than with free amine. Polyketimine: polyamine blocked with a ketone

8) **Epoxy resin** or **polyepoxy** compound (1) + **polycarboxylic acid** or **acid resin** (2):

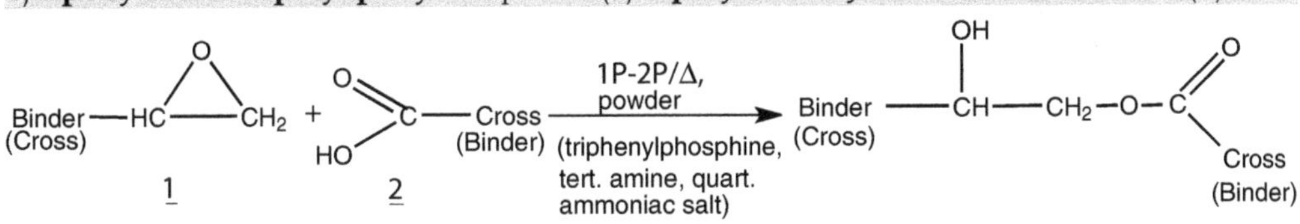

9) Epoxy resin (1) (containing OH) + **polyanhydride** (2):

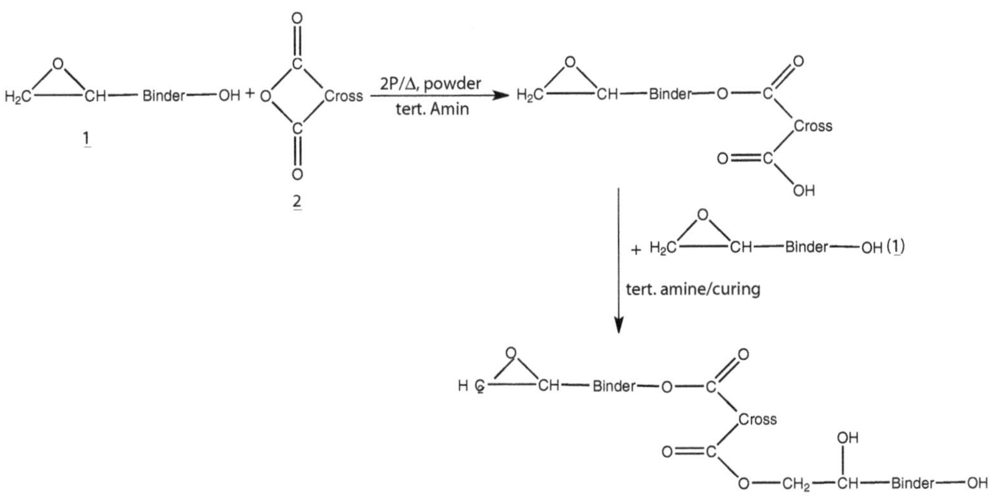

Two-step curing; stoving required for completion

10) Silicone resin (1) Self-crosslinking (condensation crosslinking):

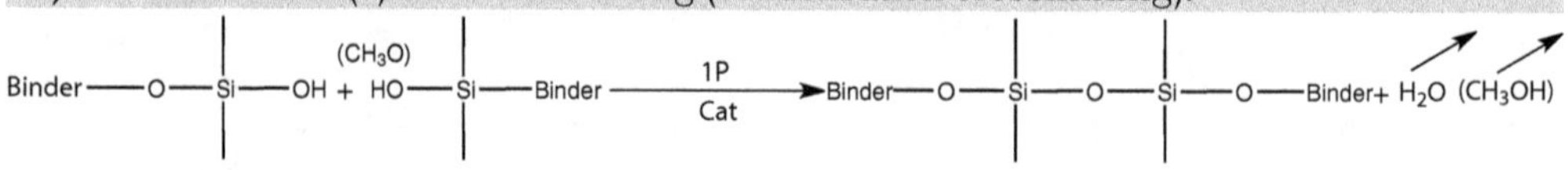

1 1

Frequently, however, co-condensation with resin polyols to yield blending resins.

11) SiH silicone resin (1) + **vinyl silicone resin** (2):

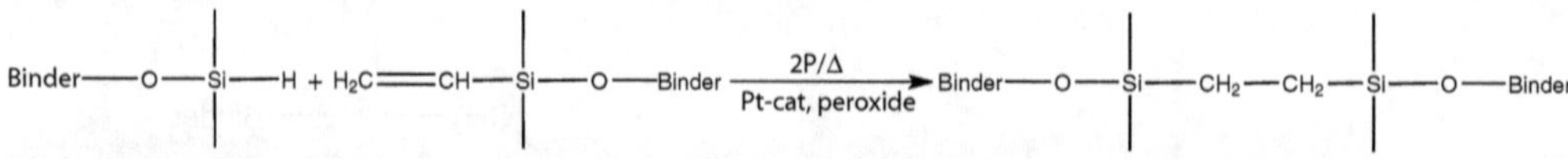

1 2

(addition crosslinking)

12) Polyacryloyl resin (3) (reactive polyacrylate) + **polyketimine** (1) with atmospheric humidity (2):

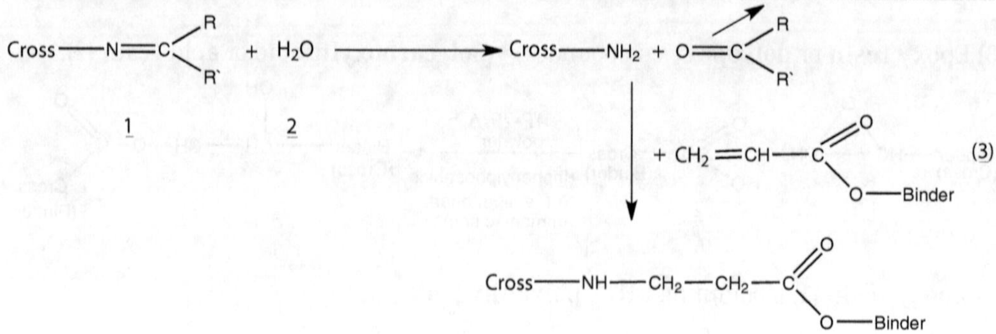

2-pack crosslinking of non-isocyanate chemistry (pseudo-Michael addition)

13) Polyacryloyl resin (1) (reactive polyacrylate) + **CH acid crosslinker** (2) (polyacetoacetate, polymalonate):

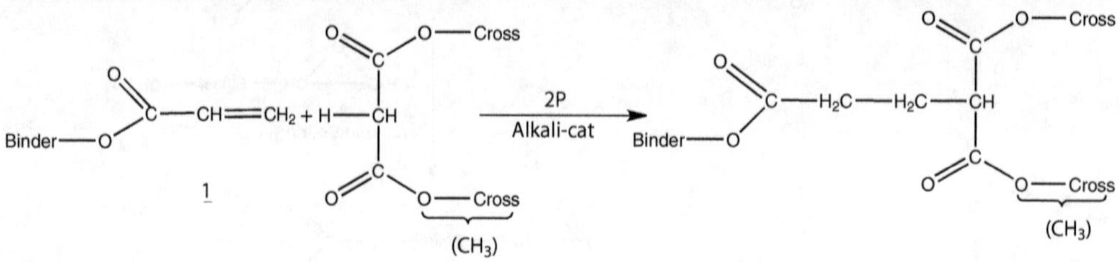

2

2-pack crosslinking of non-isocyanate chemistry (Michael addition)

14) Carboxy dispersion (1) + **epoxysilane crosslinker** (2):

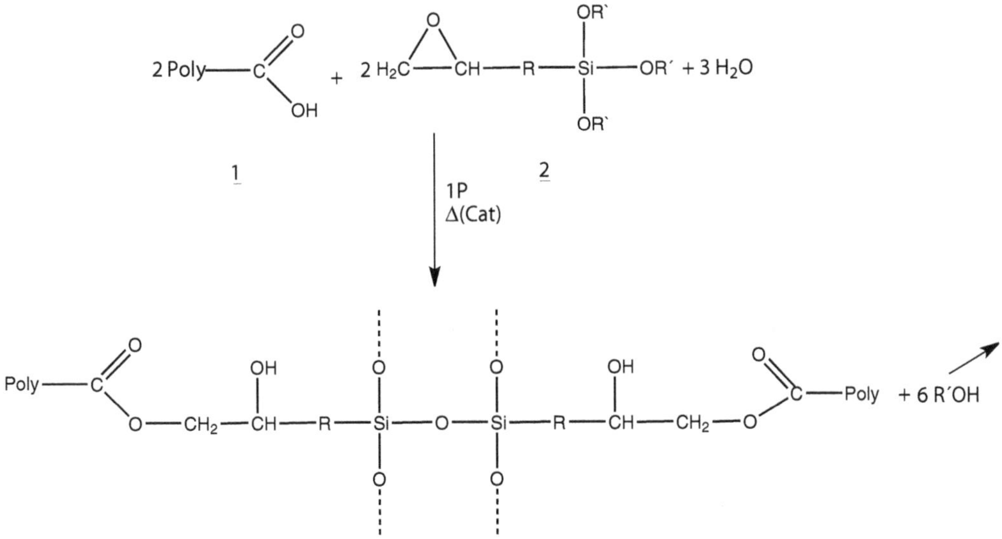

15) Carboxy dispersion (1) + **polycarbodiimide** (2):

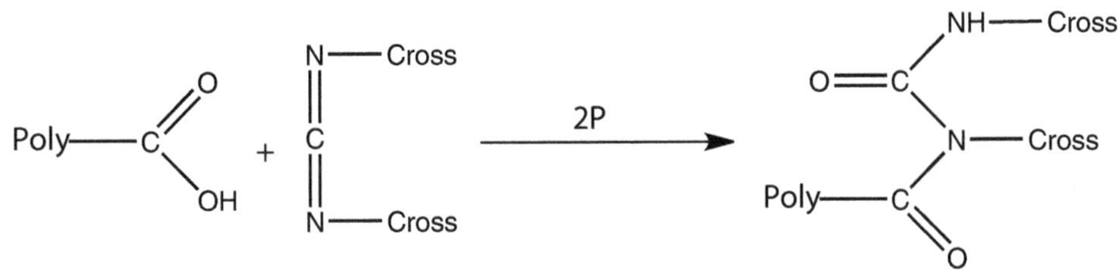

1 2

16) Carboxy dispersion (1) + **polyaziridine** (2):

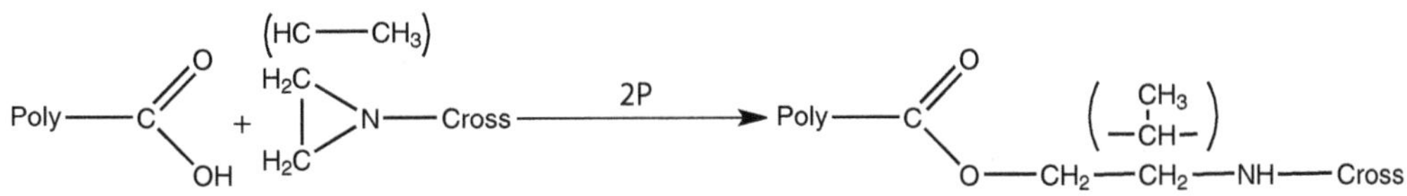

1 2

17) Carboxy dispersion (1) + **zinc** or **zirconium complex** (2):

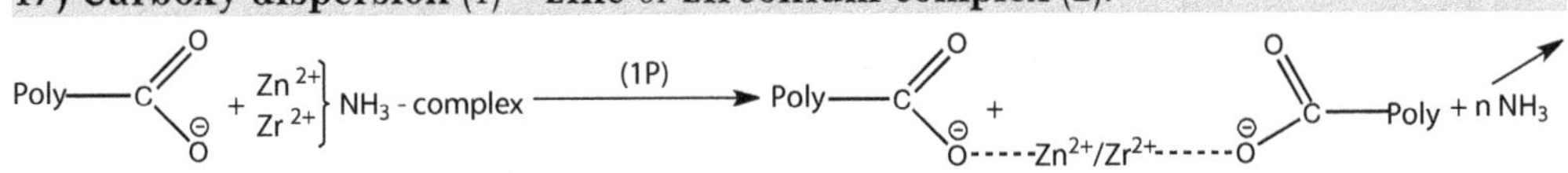

1 2

("metal salt crosslinking")

18) Carbonyl dispersion (1) + **dihydrazide** (2):

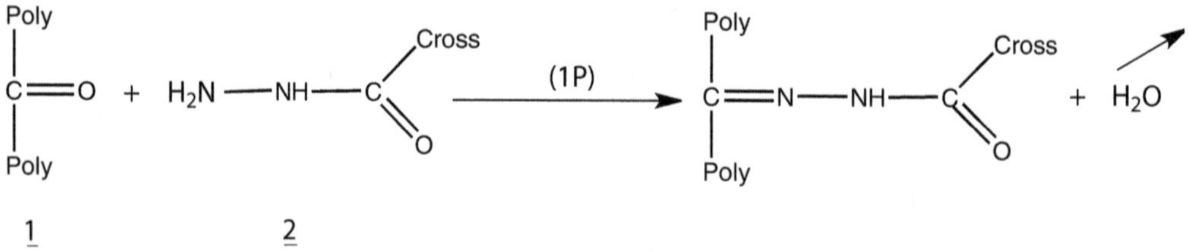

8.4 Selected crosslinking reactions in detail

8.4.1 Formation of polyurethane systems

8.4.1.1 Conventional 2-pack curing

Polyurethane formation from free polyisocyanate and polyol, in accordance with reaction 1 from Chapter 8.3, is a room-temperature polyaddition which speeds up at the end and thus can only be performed as a two-component reaction. The base resins are resin polyols and especially polyester, polyacrylopolyols, short-oil alkyd resins and longer-chain epoxy resins; for the high-end waterborne sector, there are also polyurethane dispersions and, for the construction sector, polyether polyols for primers. **Paint polyisocyanates** (hardeners) are oligomers that primarily contain three isocyanate (NCO-) groups per molecule. These are divided at the topmost level into **aromatic** and **aliphatic polyisocyanates.** The former are more reactive, but are yellow or tend to yellow extensively in light, and so are only suitable for functional coatings and primers. The latter react more sluggishly, but are light-fast and thus the hardeners of choice for demanding topcoats.

The reaction is believed to proceed by the following mechanism [22]:

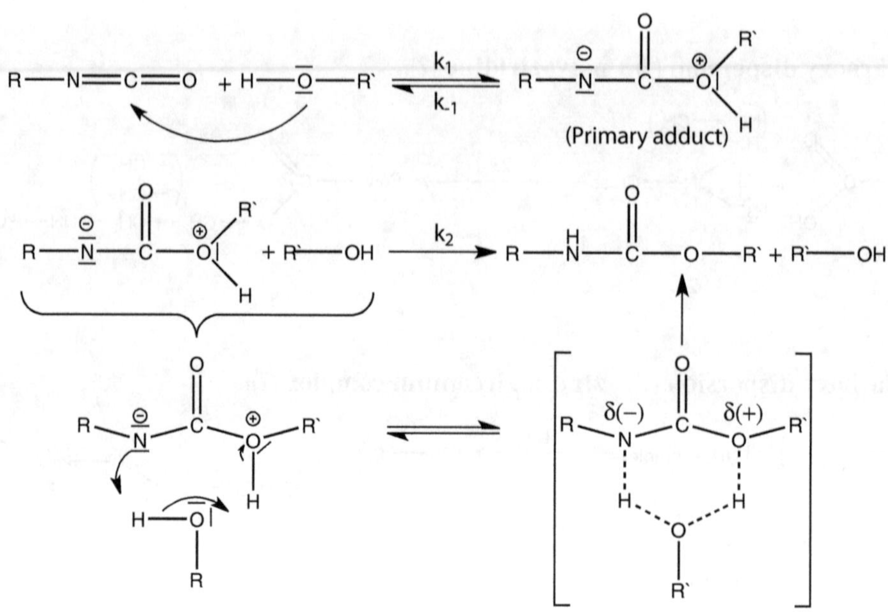

On the rough assumption that the concentration of the primary adduct is constant, the overall reaction rate is:

$$r_{tot} = k_1[NCO][OH]\frac{k_2[OH]}{k_2[OH]-k_{-1}}$$

Where the OH concentration is high, i.e. at the start of the reaction or in the 2-pack mixture, the reaction is virtually first order with respect to [OH], since k_2[OH] in the denominator is large versus k_{-1} and thus can be omitted from the fraction. At a low OH concentration, to an approximation, only k_{-1} remains in the denominator and the reaction is then second order with respect to [OH]. However, the reaction does not always follow these kinetics [22].

This fact is unfavourable for curing, because the curing rate decreases sharply as the reaction progresses. Curing with aliphatic polyisocyanate is usually catalysed with **dibutyl tin dilaurate (DBTL)** and/or a tertiary amine, such as **diazabicyclooctane (DABCO)**, which react synergistically. Apart from generally accelerating the reaction, catalysis offers the advantage of reducing the OH reaction order from two (no catalyst) to 0.5 and, in the case of DBTL, to one, when an amine catalyst is used. The outcome is a more favourable ratio of potlife to curing time [22].

The first key step in the isocyanate reaction is the addition of an electron-rich, nucleophilic group, such as -OH (in the case of polyols) or =NH (amines) to the electrophilic C atom of the NCO group to form a zwitter ion. This step proceeds all the faster if the the atom attacking the NCO group is nucleophilic (N is stronger than O) and the C atom of the NCO group is electrophilic, i.e. more positively charged. Since the C atom of the NCO group in an aromatic polyisocyanate is additionally positivised by the negative mesomeric effect,

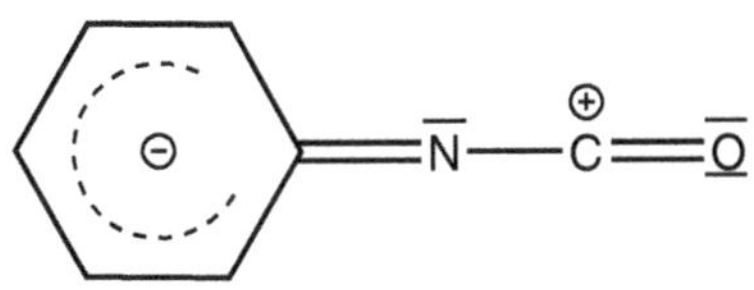

aromatic polyisocyanates are much more reactive than aliphatic ones in which such a charge delocalisation is not possible. The second step in the overall reaction, namely proton transfer, can be viewed as stabilisation of the adduct.

Non-waterborne systems are governed by the relationships between solvent evaporation (provided it is not a 100 % liquid system), rise in T_g, transition from reaction control to diffusion control and the topological conversion limit (see Chapter 8.2.2). To maximise the last of these and to obtain optimum mechanical film properties, the hydroxy groups should be dispersed as uniformly as possible along the polymer chains, a situation that does not obtain in the case of random copolymers, such as polyacrylopolyols especially, and is likely one reason for the generally superior flexibility of crosslinked **polyesterpolyols** relative to polyacrylopolyols. In the case of polyurethane systems, there are further specific chemical reasons to be taken into account as well. When NCO/OH reacts with unreacted isocyanate, water diffusing from the atmosphere is also reactive:

$$\text{R-NCO} + H_2O \longrightarrow \text{R-NH}_2 + CO_2$$

The amine which is formed then immediately reacts further with isocyanate to form a urea bridge :

$$\text{R-NCO} + \text{H}_2\text{N-R} \longrightarrow \text{R-NH-C(O)-NH-R}$$

The stoichiometry of this reaction shows that two NCO groups are consumed in the **binding of urea**. Essentially, therefore, reaction with water yields a lower crosslink density than with polyol. To compensate for this loss, it is common practice to use a certain excess of isocyanate, e.g. 10 %.

The undesired reactivity with water also means that the paint raw materials should contain as little water as possible. During application, too, excessive atmospheric humidity can cause enough water to get into the curing paint to form CO_2 which gives rise to bubbling, needling and cratering.

The rate at which urethane is formed is also determined by solvents – this is generally called the **solvent effect**. Strong **hydrogen bond acceptors**, such as esters, ketones and ethers make much more suitable solvents and diluents than, e.g., hydrocarbons, which generally immobilise protons and thus interfere with protonation of the isocyanate nitrogen [22]. This effect is beneficial to the pot-life: the solvent present in the bulk of the paint retards the reaction, but disappears from the film during drying, allowing the curing to proceed without interruption.

Curing can proceed very rapidly, despite the long pot-life, by adding the amine catalyst as vapour – e.g. along with the spray air or in a downstream chamber – during or after application of the paint (**vapour curing**).

8.4.1.2 Waterborne 2-pack curing

Probably more by luck than as the result of systematic development, it was discovered in the late 1980s that polyisocyanates can also be used as hardeners in aqueous phase. Generally, a waterborne 2-pack paint consists of a secondary or primary dispersion of a resin polyol into which a conventional (hydrophobic) or spontaneously emulsifying (hydrophilised) polyisocyanate has been emulsified. The isocyanate is present as fine droplets ranging from 0.1 to 10 μm in size that are surrounded by the generally much smaller polyol particles and thus are largely shielded from the water. Nonetheless, slowly but surely, **polyurea formation** commences immediately in the 2-pack mixture through reaction with water. The polyurethane forms by reaction with the resin polyol, as a result of which the droplets become more and more viscous. After application, the water evaporates relatively quickly, along with any amine present. The droplets coalesce and the polyol finally mixes with the polyisocyanate by diffusion [2, 90].

Essentially, the polyisocyanate is involved in three competing, parallel reactions:

- polyurethane formation with the polyol (dominant reaction)
- polyurea formation with water (subordinate to roughly just as fast)
- amide formation with the carboxy groups of the polyol as follows

$$\text{R-NCO + HOOC-R}' \longrightarrow \text{R-C(O)-NH-R}' + CO_2 \text{ (subordinate).}$$

The ratio of these reactions to each other varies substantially with the application and drying conditions. The sooner the water can leave the film – e.g. through increased temperature – the less polyurea that is formed [91]. The consumption of isocyanate in a minor reaction is offset by a corresponding isocyanate excess, which can lie between 30 and 100 %: $n_{NCO} : n_{OH} = 1.3$ to 2. The isocyanate excess is not lost, however, since it also generates network bonds, albeit less efficiently.

For practical purposes, there are two particularly important rules (the manufacturer's instructions must be obeyed, too):

- polyisocyanate must be incorporated extremely thoroughly into the polyol base paint in order that fine droplets may be formed and the components distributed homogeneously in the film
- after the potlife has expired (this generally cannot be determined from an increase in viscosity rise), the mixture may no longer be used as since the droplets in the film will no longer fully flow or diffuse into each other.

- typical data are provided below:
  - room-temperature curing [90]:
  - crosslinking ratio: n_{NCO}:n_{OH} = 1.5 to 2
  - after 30 min (from application): 95 % of volatiles have evaporated
  - after 45 min: film completely transparent
  - after 60 min: film is dust dry
  - after 2 to 3 h: film is touch dry
  - T_g after five days: room temperature
  - T_g after complete curing: approx. 40 °C
- thermal curing [92]:
  - crosslinking ratio: n_{NCO} : n_{OH} = 1.5
  - after 8 min at room temperature + 5 min at 80 °C (after application): less than 0.5 % water
  - after a further 5 min at 130 °C: less than 0.2 % water
  - residual NCO content after 8 min at room temperature + 5 min at 80 °C + 30 min at 130 °C: 7 % of the initial value
  - after a further seven days at room temperature: less than 0.2 % residual NCO (of the initial value), approx. 91 mol.% urethane, approx. 6 mol.% urea, less than 3 mol.% amide groups same paint dried completely at room temperature: approx. 51 mol.% urethane and 46 mol.% urea groups).

It has been pointed out elsewhere in the literature that, optimum formulation and processing prevent CO_2 **reaction bubbles** from occurring since the CO_2 formed by reaction with water and carboxy groups dissolves quickly in the wet film and then diffuses out of the film; any bubbles present in the film are air bubbles entrained during application. Over-rapid water evaporation after application, in contrast to what might be expected from a chemical viewpoint, can even be disadvantageous. The reason is that entrained air bubbles cannot rise quickly enough to the surface because the film is solidifying too quickly by physical means. It is therefore advisable to avoid curing in overly dry air [93].

8.4.1.3 Moisture curing

The most important class of **moisture curing polyurethane coating materials** contain NCO functional oligomers (prepolymers) by way of binder. They undergo the following reaction scheme by absorbing atmospheric humidity (see Section 8.4.1.1) to yield urea. The carbon dioxide formed thereby can lead to bubble formation if the maximum film thickness, which depends on the formulation and the application conditions, such as temperature and atmospheric humidity, is exceeded. These paints have a virtually unlimited shelf life if

$$R-NCO+H_2O \rightarrow R-NH_2+CO_2;\ R-NH_2+OCN-R \rightarrow R-NH-C(O)-NH-R$$

hermetically sealed off from air.

8.4.1.4 Stoving of blocked polyisocyanates

The desire to have not only 2-pack PUR systems, but also 1-pack PUR paints has led to the development of blocked (**terminated**) polyisocyanates. **Blocking agents** are generally substances which have a cleavable proton and so add relatively strongly, yet reversibly, to isocyanate groups. Polyisocyanates deactivated in this way are practically inert in mixtures with polyols, and possibly even in water. At elevated temperatures, an NCO/OH curing reaction occurs, in which the blocking agent is cleaved (see reaction 3 in Chapter 8.3). The usually fairly high **unblocking temperature** in excess of 150 °C can be lowered by adding DBTL, tertiary amine or other more specific unblocking catalysts [2, 94]. For blocked polyisocyanates, too, the aromatic representatives are more reactive than the aliphatic kind [22].

Of all the unblocking mechanisms which have been proposed, the highly plausible **elimination-addition mechanism** is presented below [22]:

$$\mathrm{R{-}NH{-}C(O){-}B} \underset{k_{-1}}{\overset{k_1}{\rightleftarrows}} \mathrm{R{-}NCO + B{-}H} \qquad \text{(Elimination)}$$

$$\mathrm{R{-}NCO + HO{-}R'} \overset{k_2}{\rightleftarrows} \mathrm{R{-}NH{-}C(O){-}R'} \qquad \text{(Addition)}$$

(B-H is a blocking agent)

If addition is very fast, e.g. for the reaction with amine instead of alcohol, elimination alone is rate determining because every NCO group which is formed enters into further reactions without delay; there is no appreciable build-up of NCO. The equation for the overall reaction rate is $r = k_1$ [BI] (BI means blocked isocyanate). The alcohol addition reaction, which is much slower, is co-determining and so the overall rate is then more or less given by [22]:

$$r = \frac{k_1 [BI] k_2 [OH]}{k_{-1} [BH]}$$

Moreover, a true, pre-equilibrium (more accurately: flow equilibrium) obtains, as a result of which the overall rate depends on the concentration of blocking agent as well, and thus on the rate at which it leaves the paint film.

A list of common blocking agents and their deblocking temperatures, expressed as the **initial crosslinking temperature** is presented in Table 8.1. Note that reaction between a polyisocyanate blocked with malonic acid diethyl ester and a polyol is a re-esterification reaction, with cleavage of ethanol, i.e. does not entail cleavage of malonic ester.

Interestingly, and in agreement with the mechanism discussed at the outset, the **initial unblocking temperatures** (uncatalysed) in the absence of polyol are on average 60 K higher than those with polyol [2].

A disadvantage of the blocking agents mentioned so far is that they are cleaved and, while some may remain in the film, the bulk of them escape into the oven air and contaminate

Table 8.1 Initial crosslinking temperatures (in the presence of polyol) for various blocking agents, with and without DBTL, as determined by DMA; from [2]
butanone oxime = methyl ethyl ketoxime

Blocking agent	Initial crosslinking temperature [°C]	
	with 1 % DBTL (in terms of solid resin)	without DBTL
ε-caprolactam	163	175
diethylamine	144	173
butanone oxime	137	156
1,2,4-triazole	121	147
3,5-dimethyl pyrazole	112	158
diisopropylamine	115	136
imidazole	98	123
malonic acid diethyl ester	–	104

it. For this reason, work is underway to develop cleavage-free, blocked polyisocyanates. A widely exploited way to achieve this with polyurethane paints is given in reaction 4 from Chapter 8.3, in which **polyuretdiones**, which can be viewed as **internally blocked** diisocyanates, react completely into the film (see Chapter 9.2.3 for more details).

An even more recent development is that of cyclopentanone-2-carboxy esters (CPME/CPEE) as blocking agents. They combine absence of cleavage with low viscosity and light-fastness [95].

8.4.2 Crosslinking of epoxy resins

8.4.2.1 Epoxy resins and their reactivity

Epoxy resins (EP resins) are reaction resins bearing sufficient numbers of epoxy groups (DIN 16945) for effecting curing, i.e. they contain three-member rings composed of two carbon and one oxygen atoms. By far the most important class of EP resins is that of **bisphenol-A** of general formula:

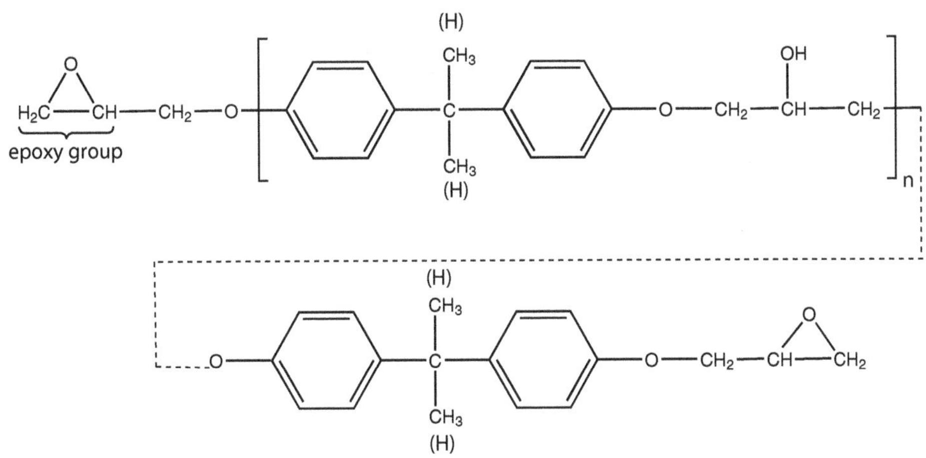

where n = 0, 1, 2, ..., approx. 26

Owing to the polydispersity of the molecular weight, the number-average values of industrial resins are not generally integers. It should also be noted that n additionally indicates the number of hydroxy groups along the molecule chain that can also or alternatively be used to crosslink these EP resins. An EP resin with a mean n value of approx. 0.14 is a liquid which is used for liquid, 100 % systems or high-solids paints. Resins having n greater than one are more or less solid and - with the exception of coating powders - are dissolved or dispersed in water for use.

There are numerous other EP resins which contain several **glycidyl groups**, as well as resins of the (cyclo)**olefin type** with **internal epoxy groups** in which the latter are located in hydrocarbon chains or in alicyclic rings:

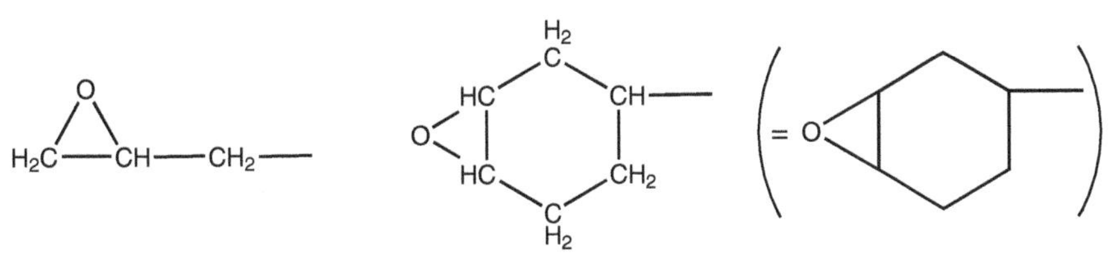

glycidyl group   EP group in alicyclic binder

Whereas bisphenol-A/-F and other polyglycidyl resins prefer to react with nucleophiles such as amines and thiols (hardeners with SH groups), the olefin types predominantly react with electrophiles, with strong acids, such as BF_3 (a Lewis acid), or with protic acids, such as photochemically generated super-acids, triggering cationic polymerisation (see Chapter 10.1.2).

The epoxy ring tries to open up under the severe internal strain of its bond angle, which is less than the tetrahedral angle of 109.5° (**valence angle strain**, **Baeyer strain**). Additionally, it is polar, with the O atom being negative (nucleophilic) and the C atoms being positive (electrophilic). Those chemical reactions of the epoxy group which are of most importance to paints are addition of the amine NH bond and addition of a carboxy group (COOH) [9, 12, 22]:

For details of cationic polymerisation, see Chapter 10.1.2.

The addition of amines is accelerated by water, alcohols, tertiary amines and weak acids, which accept protons from the N atom of the activated complex (≠). Since addition of the epoxy groups leads to the formation of OH groups, a certain degree of **autocatalysis** exists,

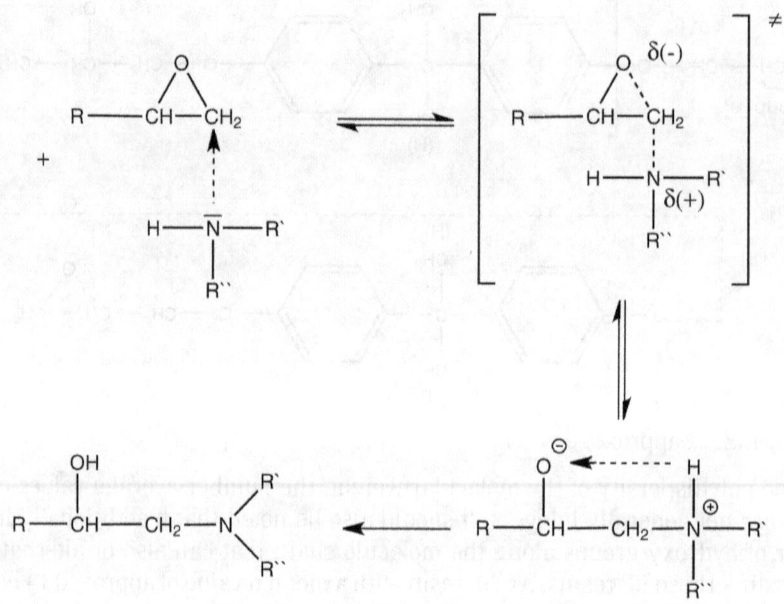

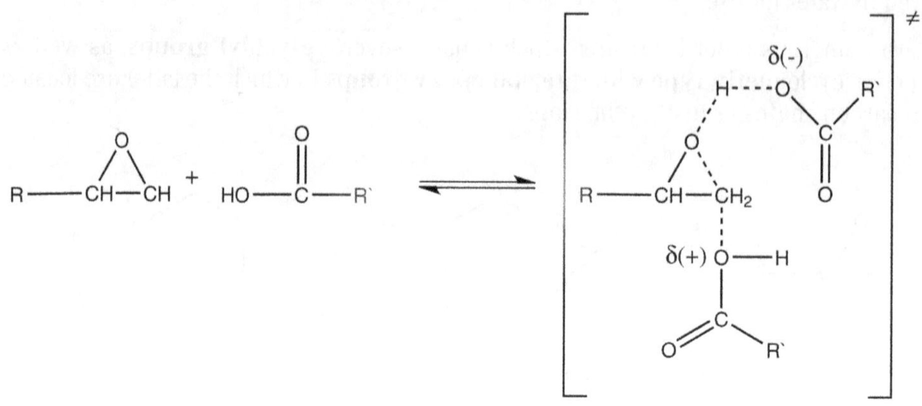

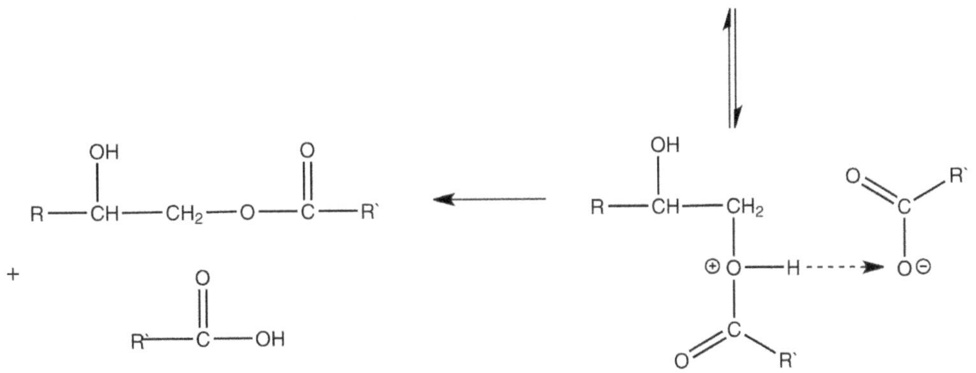

i.e. the reaction is somewhat self-accelerating. It can also be readily understood, then, why the additive **2,4,6-tris-(dimethylaminomethyl) phenol**

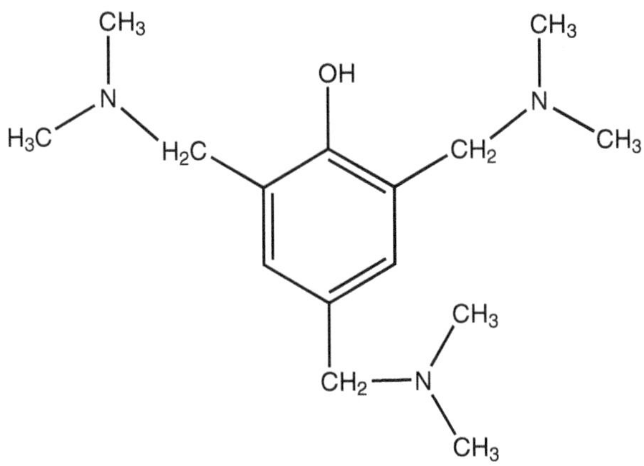

is such an effective accelerator for amine curing: both the tertiary amino groups and the phenolic (acidic) OH group catalyse the reaction [22].

The transition state in the carboxylic acid addition contains an epoxy group and two acid groups. This explains the observed third-order kinetics: r = k [EP][COOH]^2 [22]. As the conversion rate rises and the concentration of epoxy and acid decreases, the reaction slows down rapidly. The high steric hindrance of the transition state gives rise to a highly negative **activation entropy** and thus to a small value of A in Arrhenius' equation (Equation 5.21). To compensate, the exponential component [exp(-E_a/RT)] must be large, i.e. the **activation energy** (E_a) must be relatively small, a fact which necessarily implies that the reaction rate has a low dependence on the temperature. And, indeed, acid addition at room temperature is appreciably fast, even though rapid, full curing requires stoving. The reaction is accelerated by tertiary amines, quaternary ammonium or phosphonium salts and, especially, triphenyl phosphine. It is even possible to formulate 2-pack systems that crosslink at slightly elevated temperature, but these have not achieved any lasting importance [22]. The primary use of the **carboxy-epoxy reaction** in paints is to be found in powder coatings (see Chapter 9.2).

8.4.2.2 Additional information about curing with amines

Amines in the widest sense (more correctly: **amino hardeners**) comprise simple aliphatic, alicyclic and aromatic amines, **polyamidoamines, amine adducts and Mannich bases** – possibly blocked with lower ketones to yield polyketimines. To be able to form a network, the hardener molecules (assuming the usual diepoxies are present) must contain more than two NH bonds; at least two primary (-NH_2) groups but, more frequently and additionally, secondary amino groups (-NH-) generally need to be present, with secondary amino groups being less reactive than primary ones because of steric hindrance. It is generally assumed that the mixture of resin and amine has a molar ratio of EP:NH = 1:1. Moderate deviations from 1:1 crosslinking as a way to modify technical properties are possible. All amine curing reactions must involve two components; the only exception is that of polyketimines, with which 1-pack systems of limited shelf-life can be formulated.

Reactivity with amines or amino groups decreases in the following order:

- aliphatic > alicyclic > aromatic amine

To harden adequately at room temperature, the last two types require accelerators, e.g. the 2,4,6-tris-(dimethylaminomethyl) phenol outlined in Chapter 8.4.1.2 and/or heat [19]. The amine addition is accompanied here by anionic oligomerisation (polymerisation) of epoxy groups, which boosts curing [1].

Amino curing generally requires a minimum temperature of 10 °C; this can be reduced somewhat with the use of accelerators. In the case of room-temperature curing, conversion comes virtually to a standstill at 60 to 80 % due to vitrification of the film, and T_g ultimately achieves values of just 40 to 50 °C, usually only after several days or more [12, 22, 96]. If this degree of curing yields inadequate hardness and durability, the only remedy is an increase in curing temperature or subsequent (postcuring), as described in Chapter 8.2.2, to yield peak T_g values of 170 to 180 °C.

Direct air drying of lower amines often leads to **amine efflorescence**, i.e. tacky, unattractive surface spots. These stem partly from exudation of the amine to the surface, where there is then too much amine for optimum network formation, and partly from the reaction of the primary amino groups with atmospheric carbon dioxide to yield inactive **carbamates**, e.g.

$$2\,R\text{-}NH_2 + CO_2 \longrightarrow (R\text{-}NH\text{-}CO_2)^-\,(NH_3\text{-}R)^+$$

These problems can be somewhat alleviated by allowing the hardener component to cure overnight with roughly half the resin component in a closed container to yield an ***in situ* adduct** and then mixing this with the remainder to produce the finished system. Such ***in situ* adduct hardeners** or **amine adducts** are available commercially.

The potlife can be greatly extended without much increase in the curing time by blocking the primary groups of the hardener with lower ketones, such as butanone (methyl ethyl ketone). The reaction of these **polyketimines** is the two-step type of reaction 7 in Chapter 8.3:

1. Hydrolytic blocking on exposure of the film to atmospheric humidity, and
2. Addition of the released amino groups to the epoxy groups.

Although it is theoretically possible to formulate 1-pack systems, too, commercial ketimine systems are usually of the 2-pack type owing to the effort involved in drying the other paint raw materials.

8.4.2.3 Waterborne epoxy-amine systems

The binder in a waterborne epoxy-amine system is an emulsion of a liquid epoxy resin or a dispersion of a solid epoxy resin to which a solution or emulsion of an amino hardener has been added. The self-emulsifying and possibly co-solvented components are either mixed just before application and then emulsified in water or applied as a ready-to-use waterborne product.

Generally, with the exception of the competing reaction of isocyanate with water, the steps described for 2-pack PUR systems in Chapter 8.4.1.2 apply here. The hardener penetrates the 2-pack mixture by diffusing into the resin particles and, through incipient polyaddition, increases their viscosity until crosslinking just starts; this occurs especially in the outer zone, and is accompanied by an increase in particle size. After the potlife has elapsed, the end of which is usually signalled by a significant rise in viscosity, the T_g and thus the minimum film-forming temperature (MFFT) of the 2-pack dispersion are so high that a glossy, and uniformly fully-crosslinked (solid, durable) film cannot form. The potlife and curing time naturally trend in the same direction, but are not directly proportional to each other, since crosslinking during film formation is augmented by physical surface drying [14, 97].

Figure 8.5 shows the MFFT of a highly reactive 2-pack system as a function of the reaction time and the temperature of the mixture. The fresh mixture has an MFFT of approx. 12 °C, which increases more and more steeply, the higher the storage temperature is. The added "help lines" illustrate that the pot-life increases from around two to about three hours at 20 °C when the drying temperature (object temperature) is raised from 20 to 25 °C.

Basically, epoxy-resin emulsions or dispersions, like all **dispersion colloids**, do not have an unlimited shelf-life. The droplets or particles slowly grow larger until visible **phase separation** occurs. Furthermore, epoxy groups are slowly hydrolysed by water to 1,2-diols (glycols), which cannot later react with amine [98].

8.4.2.4 Additional information on curing with polyanhydrides

Curing of epoxy resins with anhydrides, such as pyromellitic acid anhydride, trimellitic acid anhydride derivatives or maleic anhydride-acrylic copolymers, (reaction 9 in Chapter 8.3), is a two-step process:

1. Ring opening of an anhydride ring by an hydroxy group to yield a semi-ester, and the release of a carboxy group.
2. Reaction between the released carboxy group and an epoxy group, with formation of ester and release of a new hydroxy group.

Since the first step proceeds noticeably at room temperature, liquid **epoxy-anhydride systems** must be processed on one hand as a 2-pack system, but on the other require a temperature of 180 to 200 °C for curing; adding amine catalysts can lower this temperature. Since every acid group released can react with just one epoxy ring, a mixing ratio of 1 mole EP groups to 1 mole anhydride rings should theoretically suffice. In practice, though, a slight epoxy excess works best [19]. The epoxy-anhydride networks are much denser than the epoxy-amine networks since, in the former case, each epoxy group can form two new bonds and the hydroxy groups along the molecule chain in higher-molecular resins can also react.

Anhydride curing in the coating sector is relatively minor and is restricted to certain powder coatings and can-interior coatings etc., where the films' good acid resistance comes to the fore.

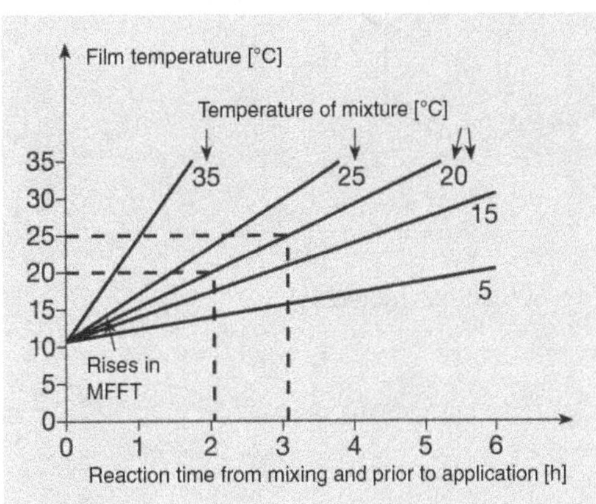

Figure 8.5: Dependence of the MFFT of a waterborne, highly reactive 2-pack epoxy-amine mixture on the reaction time and temperature of the mixture (in the pot); see text for explanation [97]

8.4.2.5 Curing via the OH groups

Long-chain bisphenol-A/-F resins, predominantly those with molecular weights exceeding 4000 g mol^{-1} ($n \geq 13$), tend to react more like polyols than like epoxy resins. They can therefore be densely crosslinked with polyisocyanates or with formaldehyde-condensation resins, such as urea, melamine and phenol resins (see Chapter 8.4.3); the films are superior to the epoxy-amine films overall. The equidistant spacing of the hydroxy groups along the molecule chains has a positive effect on flexibility. Whereas crosslinking with free aromatic polyisocyanates at temperatures as low as 0 °C is rapid and dependable (which is an advantage over epoxy-amine!), 1-pack systems with blocked polyisocyanate or formaldehyde resin must be stoved [1, 19].

A special case in this connection is the binders used for **cathodic electrocoating** (see Chapter 2.2.2). These usually consist of a basic epoxy resin-amine adduct and a blocked, aromatic polyisocyanate, which may also be chemically immobilised on the EP resin. In acid baths, the binder – just like the other paint ingredients – is dispersible, but not truly soluble and so it must be constantly stirred or circulated. When the deposited, dewatered layer is stoved at 160 to 180 °C, the isocyanate groups react with the hydroxy groups of the epoxy resin after having been unblocked.

8.4.3 Curing of resin polyols with formaldehyde condensation resins

Formaldehyde condensation resins, whose role in paint technology is to crosslink resin polyols, are formed by the transformation of nucleophilic starter molecules, such as phenol, melamine, urea etc, with a molar excess of formaldehyde (an electrophile), in the presence of lower alcohols, such as n-butanol, i-butanol, and methanol. Accordingly, these are generally referred to as **phenol resins** (PFs) or – in contrast to the non-crosslinking **novolaks** – more specifically as **resols, melamine resins** (MFs), **urea resins** (UFs) etc. The reactive groups are **methylol (hydroxymethyl)**, **alkoxymethyl groups** (etherified methylol groups) and possibly also NH bonds.

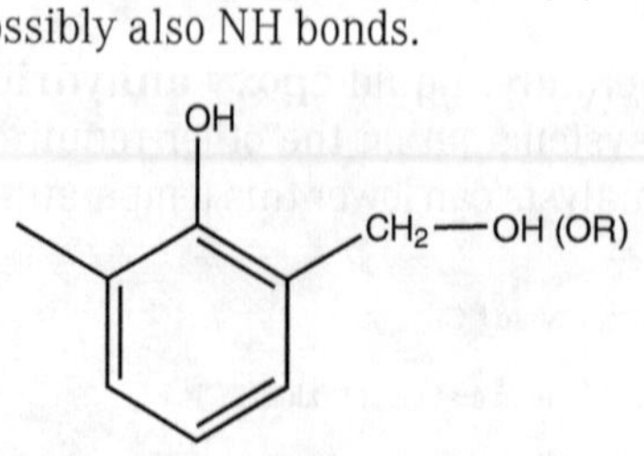

resol (part of a molecule)

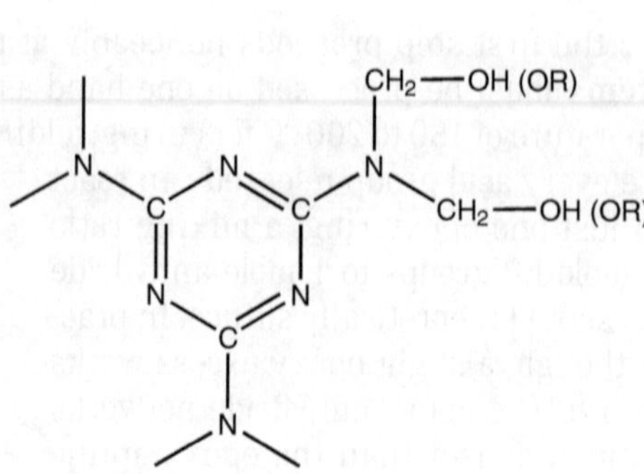

melamine (part of a molecule)

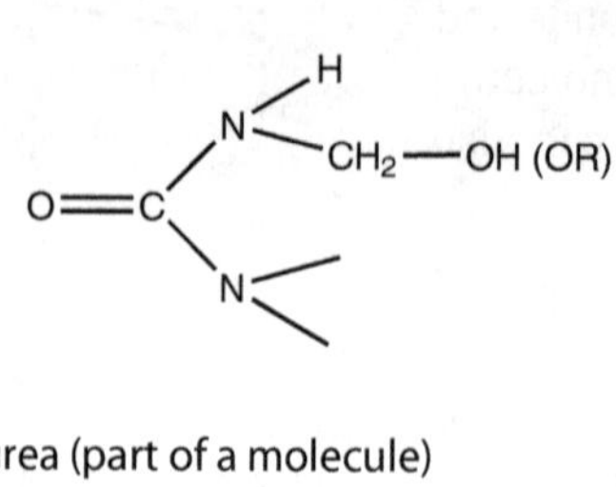

urea (part of a molecule)

$R = n\text{-}/i\text{-}C_4H_9\ (CH_3)$

The resins are relatively low molecular (oligomeric) and are commercially available as either concentrated solutions in an excess of the etherification alcohol or more or less solvent-free. During stoving, they react with a resin polyol (short-/medium-oil alkyd, saturated polyester, polyacrylic polyol, epoxy resin) via the hydroxy and carboxy groups possibly contained therein, as well as with themselves in parallel, to form a network. Reactions 1, 2 and 3 in Chapter 8.3 are the key crosslinking reactions. **Etherified methylol groups** ($-CH_2-OR$) of the crosslinker resins react with hydroxy groups much more sluggishly than do free methylol groups. As a result, the resins' **degree of etherification** can be controlled via their reactivity. Highly-etherified resins require stricter stoving conditions, but this is offset by their having a longer shelf-life, even as a 1-pack paint in combination with resin polyol.

Reaction with carboxy groups which are present in water-thinnable resin polyols due to hydrophilisation is slower than that with hydroxy groups. At the same time, the carboxy groups accelerate methylol crosslinking [9, 19].

As the degree to which **external** and **self-crosslinking** occur in parallel cannot be quantified exactly, it is not usual to calculate the stoichiometric mixing ratio for these paints. It is more common to in practice to use the mass ratio of the hard crosslinker resin (**complementary resin**) to the flexibilising polyol, with a typical ratio being 30:70.

The stoving temperature of the 1-pack systems is frequently lowered by adding an acid catalyst. Admixing several percent of a strong acid, such as para-toluenesulphonic acid, can even induce curing at room temperature. Such a system is called an acid-curing 2-pack system, in which the base paint contains the polyol and the crosslinker resin, while the "hardener" is a pure acid solution. **Acid-curing paints** consist mostly of urea and short-oil or medium-oil alkyd resin in a weight ratio of 60:40; the potlife is several days, despite rapid drying. Curing essentially consists in self-crosslinking of the urea resin. The alkyd thus largely only plays the role of an **external plasticiser** for what is essentially a very brittle urea-resin network, with oxidative drying additionally capable of leading to post-curing [9, 19]. The principal application is that of industrial wood coating. Owing to the highly undesirable **formation of formaldehyde** during curing, **acid curing** has lost a great deal of its importance.

Melamine resins differ from each other not only in their average molecular weight and viscosity, but primarily in the degrees of **methylolation** and etherification, and in the alcohol used for etherification. Consequently, the resins can be roughly grouped into the following basic types:

Type A: Highly methylolated and highly etherified; low reactivity; etherification alcohol is generally methanol: very low average molecular weight; known as **HMMM (hexakismethoxymethyl melamine)**; low viscosity and directly thinnable with water - and thus suitable for high-solids and/or waterborne paints; schematic structural formula:

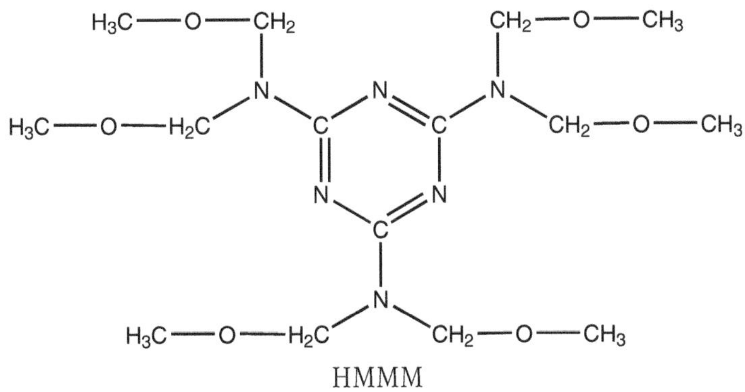

HMMM

Type B: Highly methylolated, partially etherified; moderate to high reactivity; etherification alcohol is butanol; conventional type; schematic structural formula of the monomer unit (bound by condensation to further units):

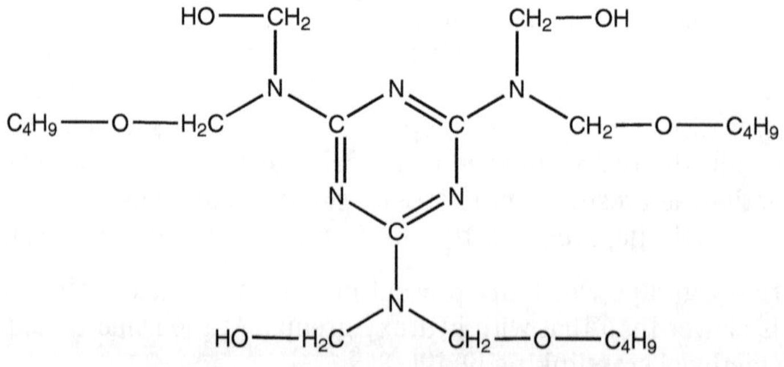

Conventional melamine resin (monomer unit)

Type C: Partially methylolated, highly etherified; high reactivity; etherification alcohol is generally methanol; little cleavage of formaldehyde during curing; schematic structural formula of the monomer unit (bound to other units by condensation):

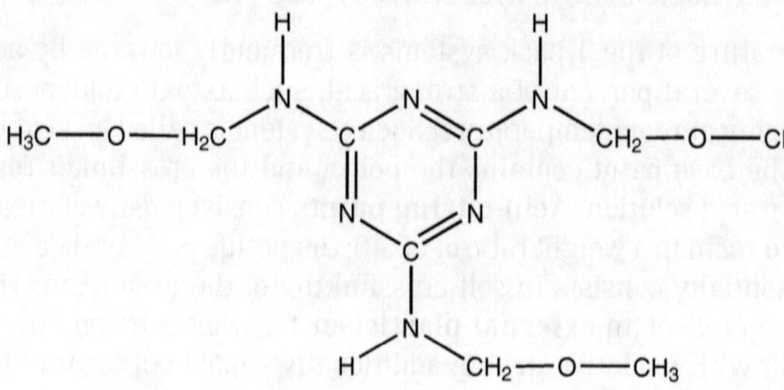

Melamine resin of the imino type (monomer unit)

Figure 8.6 shows film curing as measured by **pendulum damping (pendulum hardness)** of a melamine resin-polyester white paint cured for 20 min, as a function of curing temperature and the strength of the acid catalyst, for type A, B and C resins.

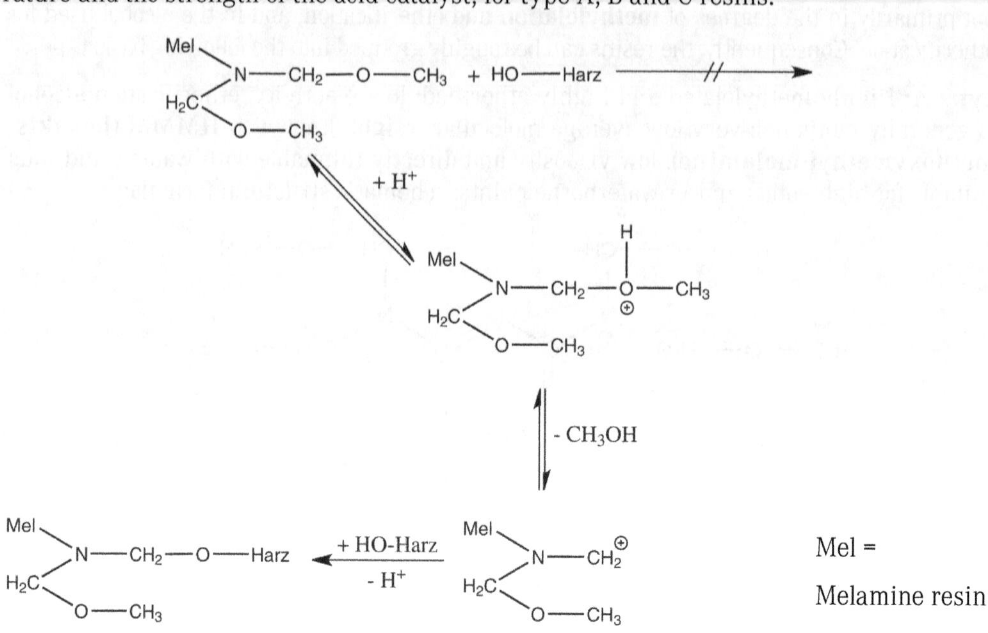

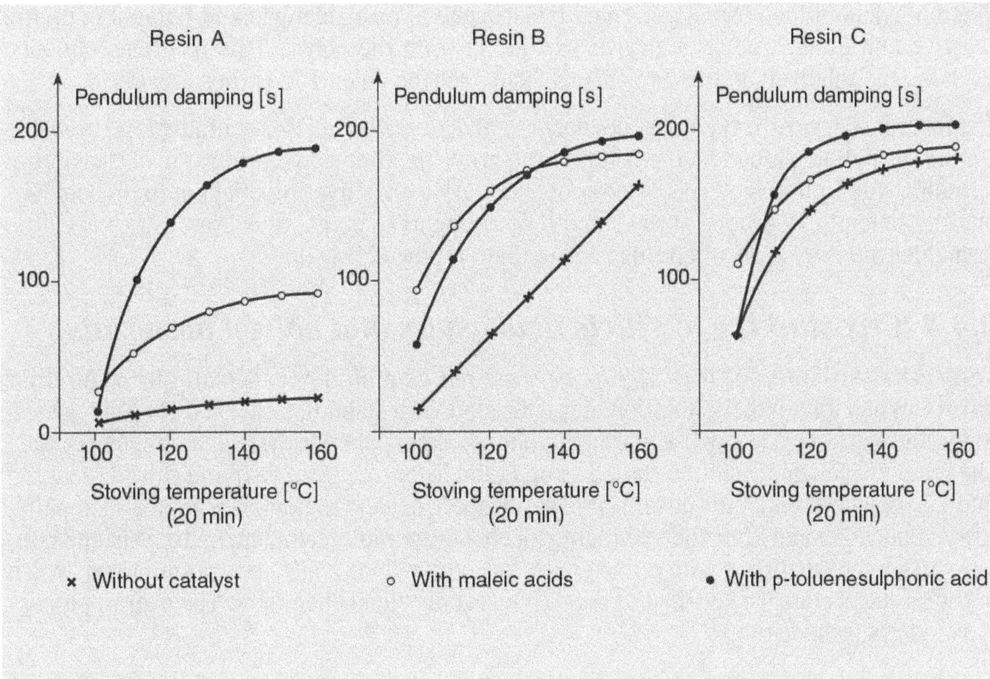

Figure 8.6: Pendulum damping as a measure of the degree of curing of three melamine resin-polyester white paints made with melamine resin types A, B and C (see text), as a function of stoving temperature and acid catalysis [9]

Resin A (HMMM type) hardens only by means of strong **acid catalysis.** Cleavage of the methoxymethyl groups, which is the rate-determining step, is needed and can only be achieved in full with a strong acid, such as para-toluenesulphonic acid (also as the blocked ammonium salt), at elevated temperature.

Appreciable self-crosslinking by the resin is only observed in the presence of water [9, 19]. The assertion in the literature that only three to four of the six methoxymethyl groups of the HMMM would react due to steric reasons, is regarded in [22] as a misinterpretation or a misunderstanding of experimental results.

The classic type B resins harden without acid – extensively so when catalysed with a carboxylic acid. Aside from self-crosslinking with formation of dimethylene ether and methylene bridges composed of two N-methylol groups, crosslinking with the polyol proceeds as shown [9, 19]:

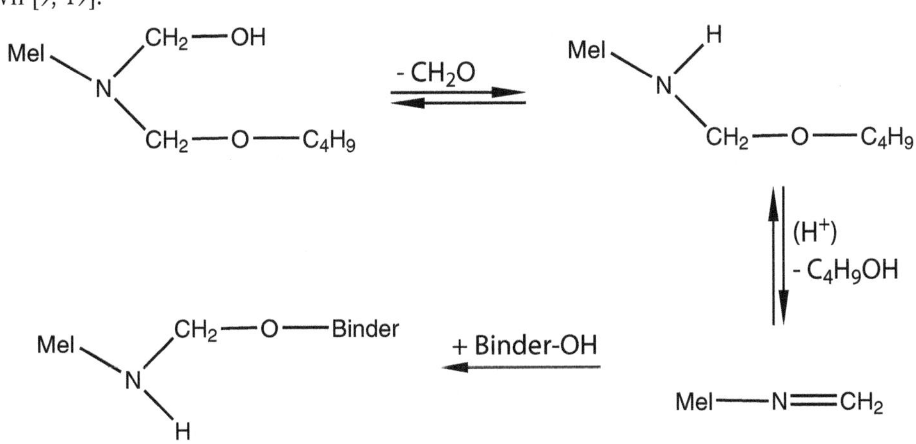

The methyleneimine (-N=CH_2) formed by cleavage of formaldehyde and butanol is highly reactive and quickly adds to an hydroxy group from the polyol. This reaction sequence explains the relatively marked release of formaldehyde on curing of these resins.

The more modern resin type C is the most reactive overall. This is essentially (and probably simplistically!) explained by the fact that the previous, time-consuming step of cleavage of formaldehyde from resin type B does not occur. Also an activating effect of hight acid NH-binding in methoxy groups at the same N-atom seems to be possible. Naturally, very little formaldehyde is evolved when this type of resin is stoved [14, 19].

8.4.4 Crosslinking of silicic acid esters and sol-gel materials

Tetraethyl silicate, $Si(OC_2H_5)_4$, has long served as a binder, especially for **zinc-dust paints**. When an ethanolic solution of it is treated with a small, judiciously chosen dose of water and a strong acid (hydrochloric), it undergoes partial hydrolysis and oligomerisation. The resultant colloidal solution, called a **sol**, is stable for roughly a year and can be rendered into a 2-pack form with zinc dust prior to application, to yield a supercritical zinc-dust paint. After application and after the stabilising alcohol evaporates, complete hydrolysis and **condensation crosslinking** with atmospheric humidity yields a SiO_2 network (quartz), which in the aqueous state is classified as a gel in terms of colloid chemistry. The overall process is as follows (schematic):

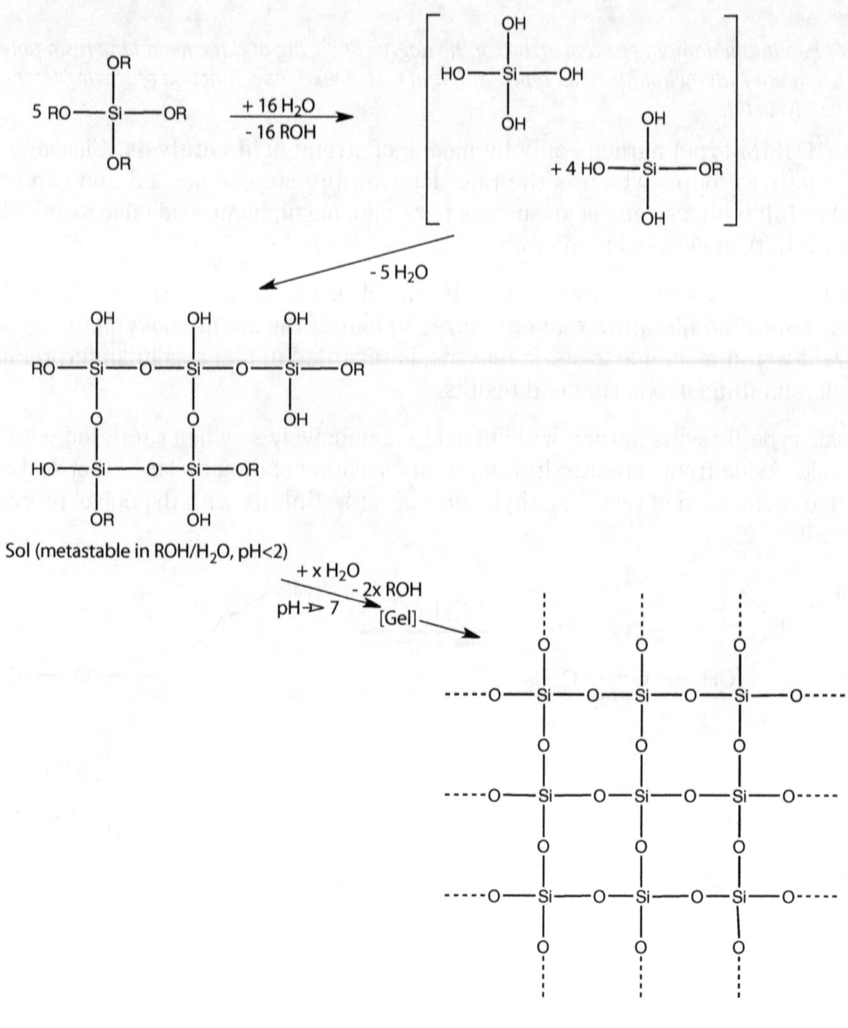

Since films obtained from pure ethyl silicate are porous and very brittle, it is more common nowadays to use a reactive **silane** of the following general structure

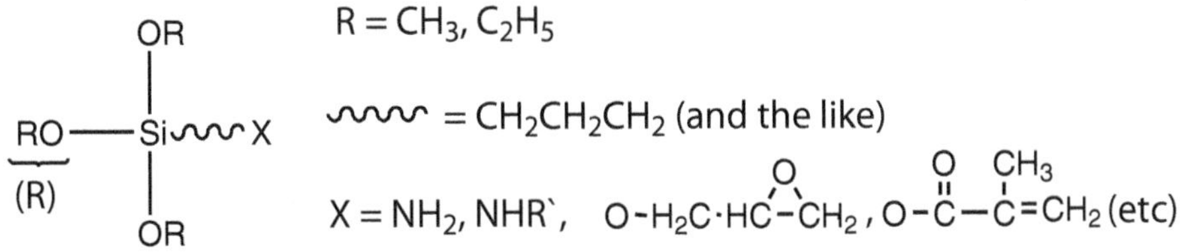

to incorporate organic groups or molecule chains, which are not hydrolysable and which are bound directly to the Si atom, into the sol. The resultant films are silkier and more paint-like, predominantly retaining their high mechanical and chemical resistance. Crosslinking reactions are also possible via the silane functional groups, which are present in addition to the crosslinkable SiOR or SiOH groups made therefrom by hydrolysis; this affords a way of integrating conventional binders to yield **hybrid binders**, which are very interesting from a paint perspective. Substituting methyl for ethyl in the SiOR groups increases sensitivity to hydrolysis and thus raises the reactivity of the silicic acid esters or silanes. The pH also plays an influential role: stability in water is a maximum at around pH 7; acid boosts catalysis of hydrolysis, while base boosts condensation. Tin compounds, too, have a catalytic effect [12]. The sols are generally applied as very low viscosity sols with high alcohol content, and yield thin wet films. Film formation is physical initially and then a continuation of the polycondensation reaction which has already occurred to a certain extent during sol formation; other reactions may occur, too. Complete dewatering and full crosslinking usually require stoving temperatures of 100 to 200 °C.

This is known as the **sol-gel process** because it involves two colloidal steps. The products nowadays would be classified as nanocomposites.

Crosslinking reaction 14 in Chapter 8.3, between carboxy dispersions and **epoxy silanes**, is very much in the mould of sol-gel chemistry [99–104].

8.4.5 General information on high solids paints

According to DIN 55945, a high solids coating material is one which has a high content of nonvolatiles. There is no generally accepted value for the minimum content of nonvolatiles, one reason being the diverse nature and quantities of the pigments of fillers contained in the various coating materials. For paints, a rough lower limit would be 65 % spray viscosity. An oft-heard expression is that of "**solids rich paints**", which strictly speaking would also include completely solventless products, such as a radiation-curable paint or a coating powder. Conversely, these paints could be called low-solvent or solvent-free paints.

Very little solvent evaporates during and after application of high solids paint. The physical consequences relative to conventional paints are

- lower decrease in layer thickness
  ⇒ better coverage of substrate texture, improved levelling
- Only minor physical surface drying or thickening
  ⇒ risk of **sagging**

During stoving, the molecular weights of the binders, which are still low initially, cause a sharp drop in viscosity that boosts the risk of sagging and edge pulling in the oven (see also Chapter 3.2-3.4).

2-Pack high-solid systems tend to have a shorter potlife and to harden more slowly than paints with a higher solvent content. Since high solid binders have low molecular weights (with a narrow molecular weight distribution) which favours a low solution viscosity, many chemical linkages have to be built up before the gel point or the finished network is produced. This necessarily entails a high concentration of reactive groups, which is additionally increased by the low degree of dilution. If second-order reaction kinetics is assumed, a doubling in concentration corresponds to a quadrupling of the reaction rate, i.e. a shortening of the potlife to around one-quarter. After the paint has been applied, although chemical curing is faster than in the case of a high solvent paint, this effect is offset by the widespread absence of physical surface drying, with the overall result that film formation is retarded [96].

Generally, the problems associated with high solid systems can be circumvented by careful formulation, especially the use of additives to modify the rheology, catalysts, and technical optimisation of the process.

9 Film formation by coating powders

9.1 General information on coating powders and how they form films

Coating powders can be divided into two groups:

- thermoplastic, and
- **thermosetting** (curing)

Thermoplastic powders, their application and film formation were covered briefly in Chapter 2.7. In view of the limited importance of this technology and the relatively simple film-forming principle, this type of coating powder will not be discussed further.

Thermosetting coating powders generally consist of angular particles (grains), in which binders, pigments/fillers as well as **deaerating** and **levelling additives**, and possibly **catalysts** are distributed as finely as possible. **Clear powders** do not contain pigments or fillers, except for possibly small quantities of mineral **matting agents**. Typical compositions (in % mass fraction) [14]:

- pigmented powder coatings:
  - binders (incl. crosslinker) approx. 60 %
  - pigment (+ filler) approx. 40 % (PVC: 15 to 20 vol.%)
  - additives therein 1 to 3 %
- powder clearcoat (acrylic for OEM):
  - binder approx. 95 %
  - additives approx. 5 %

The mean particle size is usually between 20 and 45 µm. Not only the mean particle size, but also the breadth of the particle size distribution exerts a strong influence on the application properties and film formation. Thus, an undesirable content of hyperfines of less than 10 µm boosts **agglomeration** and thus leads to impaired levelling and nibs, while overly large particles also impair levelling of the coating during stoving [7, 14]. As a rule of thumb, the average grain size should be no more than half the subsequent layer thickness [105].

In the manufacture of powder coatings, the paint ingredients spend a short period at 80 to 130 °C in an **extruder**, where the binder or the binder resin mixture is extensively homogenized in melt form with the other ingredients. These binders must not enter into any significant reaction, as otherwise molecular enlargement or incipient crosslinking would prevent the melt viscosity from dropping sufficiently during subsequent stoving. The result would be deficient levelling. As a rule of thumb:

- extrusion temperature must be at least about 50 K below the stoving temperature [22]

This means that the lowest-possible stoving temperatures (at the moment) are around 130 to 140 °C (**low-bake/low-stoving powders).** This creates a further problem: in order for the binder in the extruder to be sufficiently liquid at 80 °C, i.e. for it to flow well, it must also have a low T_g. A T_g of less than 50 °C harbours the risk that the powder particles will cake together during grinding and storage at temperatures of (in summer) up to 35 °C. Low-stoving powders therefore have to be kept extra cool during grinding, transport and storage.

For a long time, the cheaper standard coating powders could not be used with heat-sensitive substrates, such as plastics and wood products, because these could not be stoved for several minutes at the high temperatures of 160 to 200 °C. Now, however, two technical methods for coating such materials are emerging:

- Powder coatings that undergo free-radical or cationic curing by means of UV radiation are made at extruder temperatures of about 100 °C and, after application, are heated to 100 to 120 °C in one to two minutes by medium-wave IR radiation, without undergoing crosslinking or suffering from impaired levelling. Only after completion of levelling, which is not hindered here by incipient molecular enlargement, is rapid curing by means of UV radiation performed [15, 16, 106, 107]. The typical weaknesses of UV curing, such as oxygen inhibition and poor through-curing of overly thick and/or overly pigmented layers can occur here too; furthermore, the resultant networks are highly heterogeneous [12]. (For UV curing, see Chapter 10).
- Highly intense, very short-wave infrared radiation (NIR radiation, see Chapter 4.2) is used to heat a coating powder (formulated so as to be absorbent) within a few seconds to a high temperature to cure it [7, 15, 108]. If, as a rule of thumb, the reaction rate doubles for every 10 K rise in temperature, a coating powder which cures in five minutes at 180 °C would cure in $300\ s/2^5 \approx 9\ s$ at 230 °C, a figure which is consistent with reality [16].

While physical reasons dictate that UV technology is preferable for clear coatings, NIR curing is better for pigmented coatings. However, the IR absorption characteristics of these usually need to be optimized [16].

9.2 Binders and how they crosslink

The binder systems used for thermally curing powder coatings are essentially:

- epoxy resin + acid polyester: **hybrid powders**
- epoxy resin + dicyanodiamide (or anhydride or phenol derivative): **epoxy powders**
- acid polyester + hydroxyalkylamide (or di/triepoxy) } **polyester powders**
- polyester polyol + blocked polyisocyanate (or amino resin) } **polyester powders**
- epoxy acrylic resin + dicarboxylic acid } **acrylate powders**
- polyacrylopolyol + blocked polyisocyanate or amino resin } **acrylate powders**

9.2.1 Hybrid powders

Hybrid coating powders are composed of solid epoxy resins of the type bisphenol-A epichlorohydrin (see Chapter 8.4.2.1) and carboxy, saturated polyesters (**acid polyesters**). The average molecular weights are a few thousand at most. Owing to their low price, straightforward paint production and universality, they are the highest-volume binder employed in the powder coating sector, accounting for approx. 50 % of tonnage. Because weathering causes **chalking**[20] of the epoxy resin content, application is restricted to indoors.

Curing is usually performed at 60 to 200 °C, the exact temperature depending on the catalyst or accelerator and stoving duration, and mostly proceeds via the **carboxy-epoxy reaction** (see Chapter 8.3, reaction 8, and Chapter 8.4.2.1). The accelerators employed may be **imidazolines** or **imidazoles** [109]. Minor reactions that also contribute to crosslinking are cationic polymerisation (oligomerisation) of the epoxy resin and esterification [14, 22].

[20] Chalking is the sunlight-induced degradation of binder that causes pigment particles to be released at the surface.

9.2.2 *Epoxy powders*

Epoxy resins will react with liquid or dissolved amines at room temperature. **Dicyanodiamide** (DICY, DCD) and modified dicyanediamide

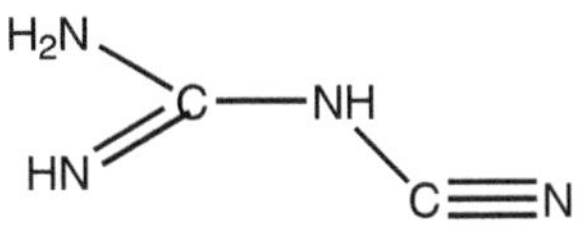

dicyanodiamide

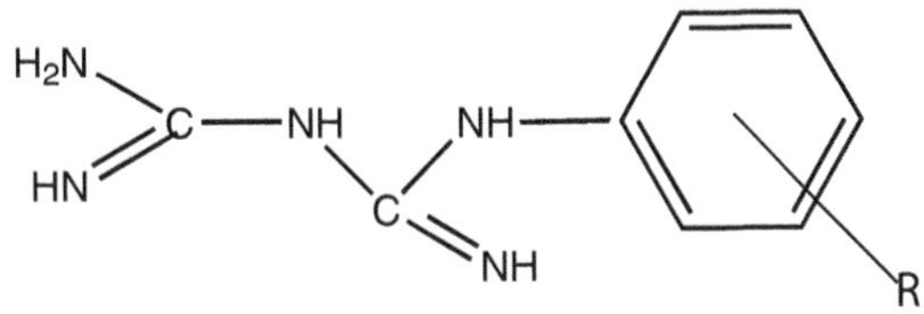

modified dicyanodiamide (ring also hydrogenated)

are amino curing agents, however, which are distributed micro-heterogeneously in the coating powder because of their high melting points of about 140 °C (if modified) to 210 °C (unmodified) and do not react in that state. During stoving, the reaction only sets in when the dicyanodiamide in the epoxy resin melt goes into solution, which necessitates stoving temperatures of 150 to 200 °C. Modification not only lowers the melting point of the pure curing agent, but also improves its solubility in the epoxy resin, and so leads to more homogeneous crosslinking [22].

The reactions between the dicyanodiamide and the epoxy resin are complex and not yet fully elucidated, but the addition of NH bonds to epoxy groups can be viewed as the principal reaction; there is also some auto-crosslinking by the epoxy resin. 2-Methylimidazole often serves as accelerator [12, 14, 22].

Crosslinking of epoxy resins with **polyanhydrides** corresponds to reaction 9 in Chapter 8.3; see also Chapter 8.4.2.3. Polyphenols (phenolic crosslinkers) react with epoxy resins by adding the acidic phenolic hydroxy groups to the epoxy rings.

Epoxy powder coatings are generally characterised by high crosslinking densities, very good adhesion to metal, good corrosion resistance and high chemical resistance. However, they are not weatherable.

9.2.3 *Polyester powders*

Carboxy (acid) polyesters can be made to react with suitable difunctional or higher-functional **glycidyl esters** instead of the non-light-fast standard epoxy resins to yield weatherable coatings. **Triglycidyl isocyanurate (TGIC)**, which was technically ideal for this and was used for a long time, was abandoned in 1998 because of its mutagenicity and has been replaced by other crosslinkers. Today two alternative types of substance are used [7, 14]:

- di- or triglycidyl esters of terephthalic and isophthalic acid, tetrahydrophthalic acid and trimellitic acid (e.g. "Araldite 910")
- **β-hydroxyalkyl amides** (tetrakis-(β-hydroxyalkyl) adipodiamides, e.g. "Primid"):

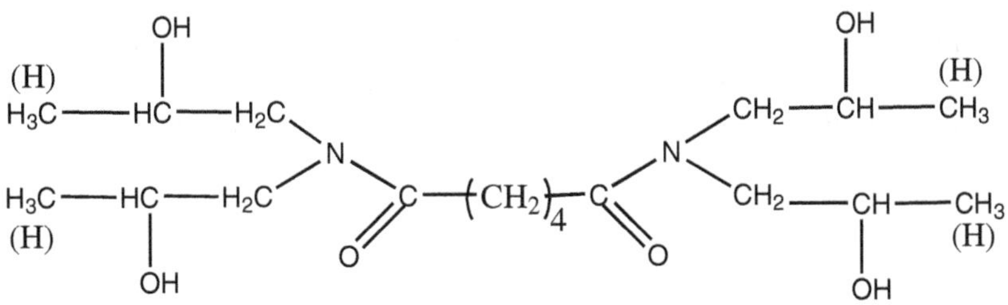

The glycidyl crosslinkers react – just like the standard epoxy resins in the hybrid powders – via the carboxy-epoxy reaction. This is a polyaddition and has the advantage of not releasing any volatile products.

Whereas carboxy groups hardly react at all with the usual alcoholic hydroxy groups at normal stoving temperatures, they form esters with hydroxyalkyl amides at 150 to 200 °C, with the reaction going to completion. The mechanism is assumed to be [110]:

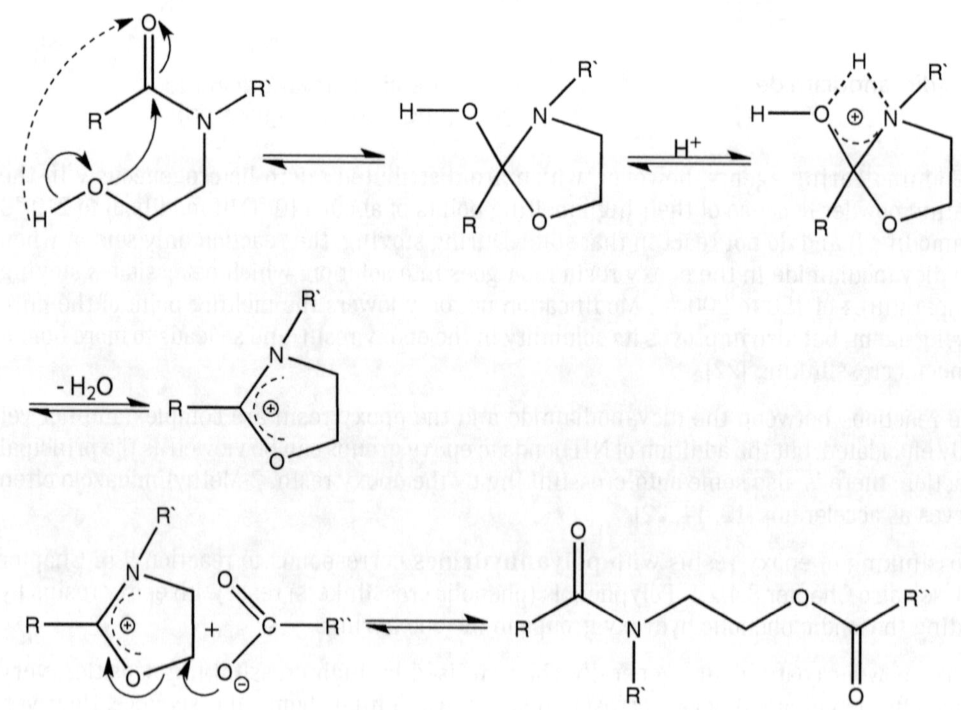

If **pinholes** or craters are to be prevented, all of the cleaved water of reaction must escape from the film (as vapour) before film levelling comes to a standstill. This means that the layer thickness is limited by the paint composition and the stoving conditions; a guide figure for this is 100 µm [111]. Apart from having to cope with this restriction, coatings crosslinked with hydroxyalkyl amides have properties comparable to those of TGIC systems [14, 15].

Hydroxy polyesters (polyester polyols) are primarily crosslinked with blocked polyisocyanates for use as coating powders (see reactions 3 and 4 in Chapter 8.3 and Chapter 8.4.1.4). Foremost among these are polyol adducts or isocyanurates of isophorone diisocyanate (IPDI), whose free isocyanate groups are blocked with **ε-caprolactam**. During stoving at 160 to 180 °C, some 80 % of the ε-caprolactam is released into the oven air. It acts somewhat as a **flux** or plasticiser in the paint film, i.e. improves levelling [12, 14, 15, 111].

Undesirable cleavage of blocking agent is eliminated by using the NCO groups of the polyuretdiones to internally "block each other." Stoving causes dissociation of the polyuretdione (1) into free polyisocyanate; this is followed by reaction with the polyol (2) to form polyurethane (4), with **polyallophanate** (3) formed by way of intermediate step:

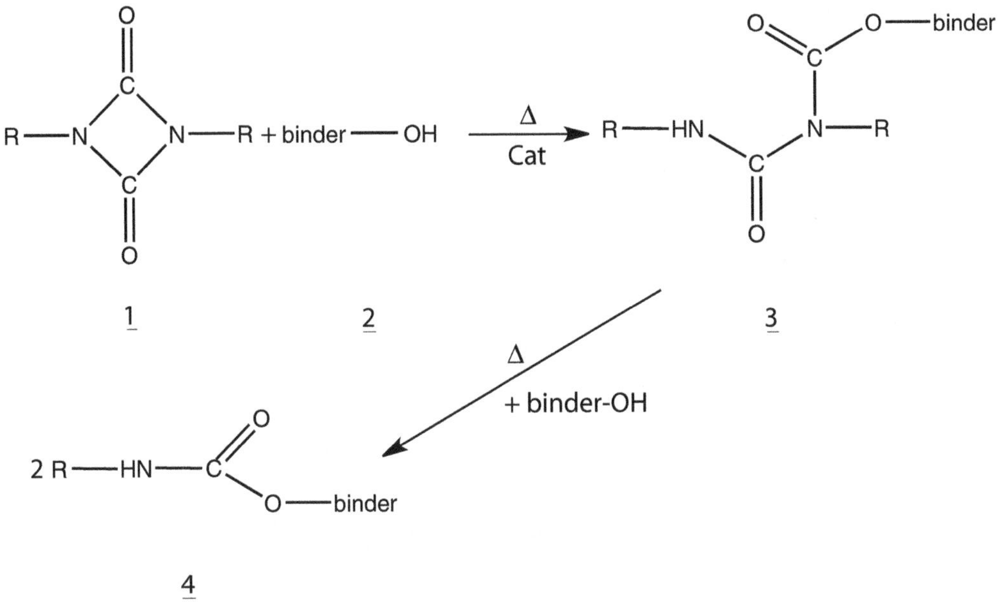

Some isocyanurate may also form as a minor product. The recent advent of special catalysts now makes it possible to lower the originally very high stoving temperature of at least 180 °C to below 150 °C [112].

The amino resin used for polyol crosslinking in the powder coating sector is glycoluril-formaldehyde resin (**glycoluril resin**) etherified with methanol. Since methanol is cleaved during curing, there can be problems with film formation. This needs to be taken into account with regard to formulation, layer thickness and stoving [22].

9.2.4 Acrylic powders

Acrylic powders are based on relatively low molecular weight, epoxy, carboxy or hydroxy **polyacrylates (acrylic resins)**. These can be crosslinked via the aforementioned reactions with dicarboxylic or polycarboxylic acids (acidic resins), polyanhydrides, polyepoxides, hydroxyalkyl amides, or polyol crosslinkers, such as blocked polyisocyanates or glycoluril resins. Combinations of different curing reactions and polymer types, such as polyacrylate and polyester (acrylic polyester hybrids), are possible, too [14, 15].

Acrylic powders have gained entry to the automotive field in particular, on account of their excellent weathering resistance, low stoving temperatures (e.g. 140 °C for 30 min), very good levelling and good optical properties [15, 114]. A major reason for their slow spread is their total incompatibility with non-acrylic powders, which required and requires the strict separation of equipment during production and processing. This problem has apparently been resolved to some extent [15].

9.2.5 General information on gelation and crosslinking

Chapter 8.2.1 covered the general principles of molecular enlargement, gelation and network formation. When the *Carothers*, and *Flory* and *Stockmeyer* equations presented there for calculating the gel point are applied without further ado to typical binder systems for coating powders, it is found that the gel points are reached only at high conversion, if at all.

- Example 1: Equimolar ratio of epoxy resin of average functionality 1.9 and acid polyester of average functionality 2.85 ($\bar{M}_n$ = 2000 g mol^{-1}, acid number 80)
  $\Rightarrow U_G = 0.77 - 0.88$
- Example 2: Equimolar ratio of ß-hydroxyalkyl amide ("Primid") and acid polyester of mean functionality 1.25 ($\bar{M}_n$ = 2000 g mol^{-1}, acid number 35)
  $\Rightarrow U_G = 1.05 - 1.16$ (no gelation, no crosslinking!)
- Example 3: Equimolar ratio of ß-hydroxyalkyl amide ("Primid") and acid polyester of mean functionality 1.87 ($\bar{M}_n$ = 3,000 g mol^{-1}, acid number 35)
  $\Rightarrow U_G = 0.62 - 0.78$

The resultant crosslink densities, therefore, are relatively low generally. This explains the moderate (if not poor) solvent resistance exhibited by many powder coatings.

In principle, denser networks could be achieved by increasing the functionality (of the base resins), but this would cause the viscosity to increase faster and the gel point to be attained earlier during curing, which in turn would impede levelling (see Chapter 9.3).

9.3 Physico-chemical aspects of film formation

9.3.1 Physical principles

After the powder has been applied to the substrate in accordance with the principles described in Chapter 2.7.1, it adheres electrostatically as an aerated layer with a surface whose flatness is governed by the shape and dimensions of the particles and particle agglomerates, along with electrostatic effects. Subsequent film formation takes the form of the overlapping individual processes of

- coalescence: fusing and flowing together of the particles during or after melting
- levelling: smoothing of the surface structure (waviness)
- crosslinking: formation of a three-dimensional polymer network

Figure 9.1 shows optical micrographs of a powder coating taken before and during levelling. Once levelling was complete, there was no more texture to be seen.

Unimpaired formation naturally requires that the substrate be properly wetted during coalescence, and that in turn necessitates careful pretreatment.

When heated at a rate of 10 to 30 K min^{-1} in a convection type oven or to over 100 K min^{-1} with conventional IR heating, the paint particles soften and then melt[21]. In this regard, the flow and diffusion processes resemble those which occur when dispersions form films

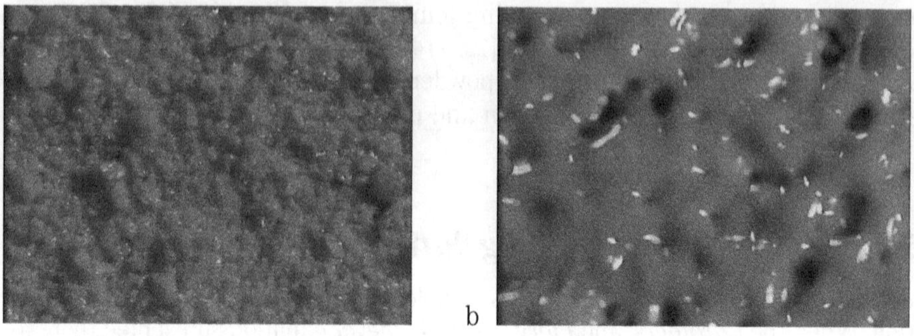

a b

Figure 9.1: a) Hybrid powder coating after application, b) during levelling (180 °C circulating air)

[21] Amorphous polymers actually do not melt; instead, there is a diffuse transition to the flowable state

(see Chapter 6.2). However, there is one key difference: there is only air here between the particles or droplets, and so there are no capillary forces generated.

The melt droplets are much less viscous than the dispersion particles, while considerable surface tension of 20 to 40 mN m^{-1} exists at the air-melt interfaces [114].

Instead of the **Frenkel equation,** which is often quoted in the literature for the coalescence of two droplets, the more generally valid semi-quantitative equation of *Nix* and *Dodge* is presented below [114]; it is also easy to deduce from dimensional analysis:

Equation 9.1:

$$t=f\left(\frac{\eta R}{\sigma}\right)$$

t is the time (corresponds to the duration of coalescence)
η is the dynamic viscosity of the droplets
σ is the surface tension of the droplets
R is the mean radius of curvature of the droplet surface
f is the functional relation (increasing monotonically)

In other words, coalescence slows down with increase in viscosity and speeds up with increase in surface tension. This is plausible, because the driving force for coalescence is the surface tension, and viscosity slows down that process.

After the droplets have coalesced completely, the film is still highly uneven, and levelling starts, basically as described by the Orchard equation (Equation 3.3). For rapid levelling, the surface tension σ must be high. However, since it changes little relative to the viscosity and in a complex way during curing (it decreases with increase in temperature and increases with increase in molecular weight [114]), it can be more or less viewed as a constant [115]. This leaves the **fluidity** integral as the factor determining levelling at a given wavelength λ and layer thickness x:

$$\int_0^t \frac{dt}{\eta(t)}$$

During curing, the viscosity passes through a minimum, which is the outcome of two opposing effects:

- during heating, the viscosity drops sharply due to dissolution of the secondary valence bonds between the molecules and possibly melting of crystallites
- as the curing temperature is approached, the viscosity rises rapidly due to incipient molecular enlargement

The initial drop in viscosity can be described by the WLF equation (Equation 5.7) or Arrhenius' equation (Equation 2.1). Since the former applies only up to a maximum of 100 K above T_g, Arrhenius' equation is proposed here because it offers a better approximation [116].

For the viscosity rise due to molecular weight enlargement, a double-logarithmic relation $\lg \eta = a \cdot \lg \bar{M}_n + b$ can be assumed [114, 116].

The shape of the viscosity curve against time, i.e. especially the steepness of its fall and rise, the depth of the minimum and the length of time it remains at low values (before rising again), is crucial to the quality of film levelling, and also to degassing (see Chapter 9.3.2). Generally, shorter heating times (higher heating rates) lead to lower viscosity minima and thus better levelling, because the physical drop in viscosity occurs faster than the chemical rise in viscosity [117]. High reactivities or (overly) fast incipient curing reactions therefore

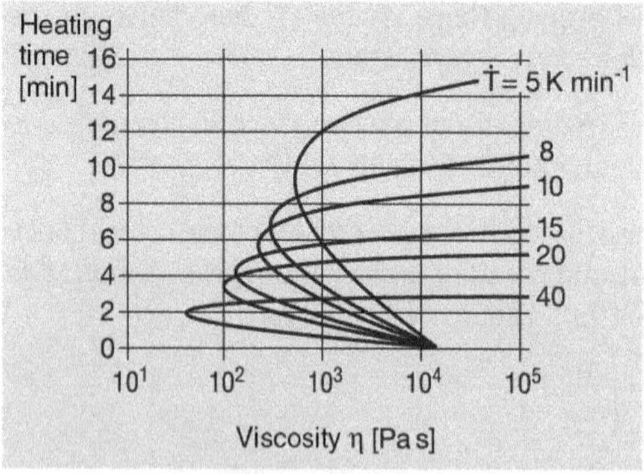

Figure 9.2: Relationship between the viscosity of a powder-coating film and the heating rate; from [117]

impede levelling. Nor is late onset of the viscosity rise desirable, since it promotes sagging and edge pulling. Figure 9.2 shows the relationship between heating rate and viscosity.

If it is assumed that appreciable levelling occurs only below 1000 Pa s, "**flow windows**" can be defined, but these are merely illustrative in nature:

- narrow and deep flow windows imply a better levelling (and less low flow) than less deep and accordingly wider ones.

The fluidity integral can be used to derive a quantitative approximation for the (theoretical) quality of levelling, called the **viscosity characteristic** [115]:

Equation 9.2:

$$VC = \int_{t_1}^{t_2} \frac{dt}{\eta(t)}$$

VC is the viscosity characteristic (for the flow limit 1000 Pa s)
t_1 is the time at which the viscosity falls below 1000 Pa s
t_2 is the time at which the viscosity exceeds 1000 Pa s
$\eta(t)$ is the viscosity-time function

Since the change in melt viscosity over time cannot yet be measured under actual stoving conditions, the flow windows are calculated from physicochemical simulation models proposed in [115].

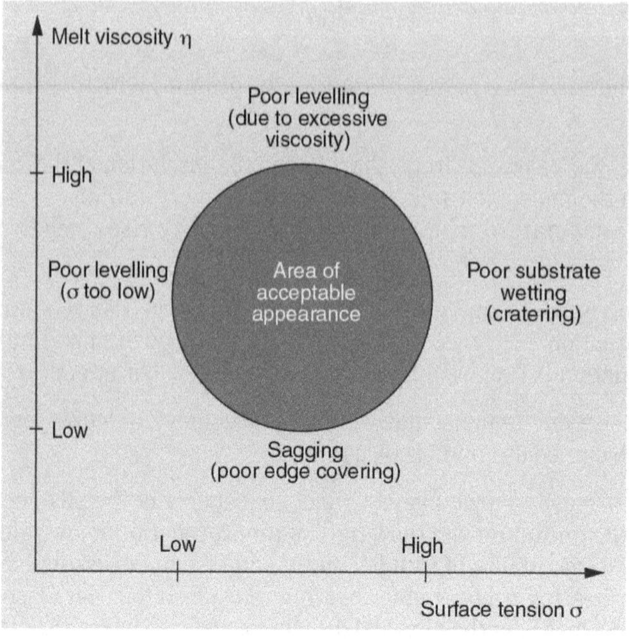

Figure 9.3: Quality of a powder coating as a function of surface tension and melt viscosity; from [3]

The following minor model calculation is likely to be of interest:

- the viscosity-time curves are modelled by parabolas $\eta = a\,t^2 + \eta_{min}$
- $\eta_{min,1} = 20$ Pa s, $\eta_{min,2} = 5$ Pa s (minimum of the second curve is lower)
- equivalent area of the flow windows and $(t_2 - t_1)_1 = 420$ s (chosen):

  $\Rightarrow \quad (t_2 - t_1)_2 = 414$ s and $\quad VC_1 : VC_2 \approx 2$

Despite the seemingly small difference between the viscosity-time curves, curve 2 should yield significantly better levelling.

Levelling of the long-wave surface unevenness of powder coatings remains a critical point, and is harder to achieve than in liquid coatings [12]. Powder coatings in some minds are still associated with orange peel texture

Finally, Figure 9.3 is a schematic diagram in which the surface quality is shown as a function of melt viscosity and the surface tension of a powder coating.

9.3.2 Effect of additives, matting

Coating powders can or must – as determined by the application area or required property range – contain all kinds of additives. For unhindered film formation, a **degassing agent** is essential; this is almost always accompanied by a **flow improver** and a **curing accelerator** or catalyst, although this can also be regarded as a binder component.

When the powder particles coalesce, some of the air in the layer becomes trapped in the form of small bubbles that slowly migrate to the film surface. If this migration proceeds too slowly, incipient solidification of the melt gives rise to thin channels in the layer that are known as **pinholes**. The most commonly employed degassing agent is **benzoin**, a solid which melts at 133 °C.

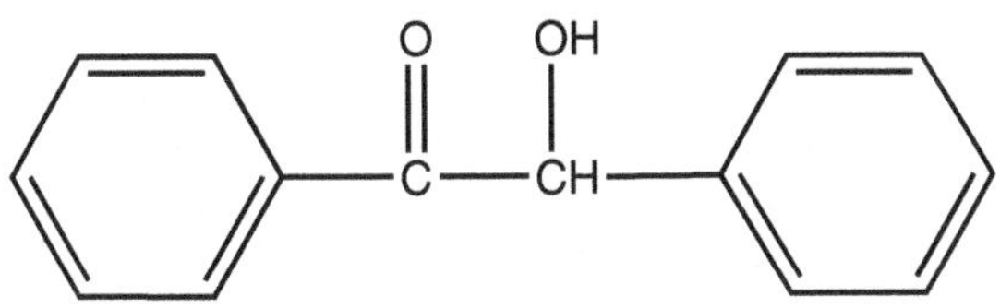

Although the mechanism of degassing is not yet fully understood, a reduction in both surface tension and viscosity of the melt was observed in the case of an epoxy resin melt [114]. Possibly, the oxidisability of benzoin to benzil plays a role.

The most important group of levelling and anti-cratering additives are relatively low-molecular and non-polar polyacrylates such as **polybutyl** and **polyoctyl acrylate**, which have poor compatibility with the paint melt and so accumulate on the surface in a uniform layer. There, they lower and – more importantly – equalise the surface tension [111, 114]. The outcome is prevention of material flows in the surface that stem from differences in surface tension and which lead to unevenness and craters.

Powder coatings are inherently glossy unless they are rendered matt. Matting of solvent-borne or waterborne paints is usually straightforward. Either the matt effect occurs of its own accord on account of a high filler loading, or it is achieved by adding matting agents, such as precipitated silicas and/or waxes. Powder coatings (and liquid 100 % systems and high-solids materials) undergo little **volume shrinkage** during film formation, however, and the matting particles or structures are not forced sufficiently towards or through the surface; the result is often poor light scattering on the surface.

Nonetheless, powder coatings can now be matted with optimized products derived from fillers, silicas, waxes and other incompatible classes of substance [111, 118–120]. The presence of **semicrystalline polyester** in addition to the amorphous kind produces matting in PUR powder coatings [121].

The aforementioned purely physically active substances yield matting results which are often mediocre and not always highly reproducible. Consequently, alternative matting concepts have been developed – particularly for extensive matting – that resort to polymer chemical mechanisms. A common principle of these methods is that one portion of the binder system reacts much faster than the rest, leading to the formation of a polymer matrix (backbone structure) which is incompatible with the portion which hardens later.

The result is micro-inhomogeneity, which manifests itself on the surface as matting roughness. The most effective variant utilises a combination of two completely different curing mechanisms.

Example:
A matt hardener of general structure

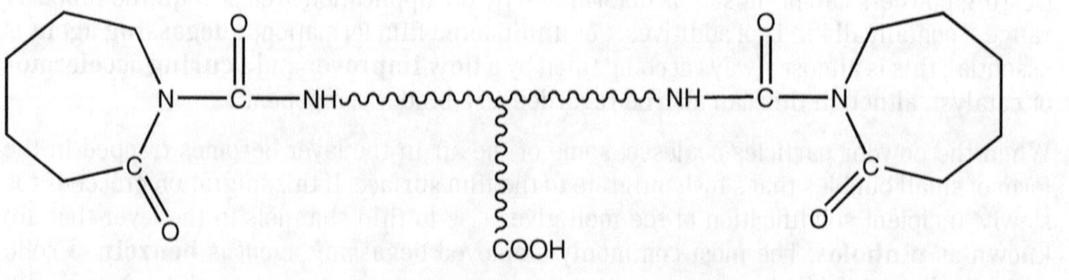

reacts quickly at 180 °C with ß-hydroxyalkyl amide or epoxy resin via its carboxy group and only then with the polyester polyol also contained in the paint, via the blocked isocyanate groups [118]. Figure 9.4 shows the heterogeneous structures in the paint film that form during crosslinking with this matt hardener; Figure 9.5 shows force scanning electron micrograph of the dull matt surface. matt hardener are also available for other types of powder coatings, e.g. amino salts of very highly functional polycarboxylic acids for epoxy resin-based powders [121, 122].

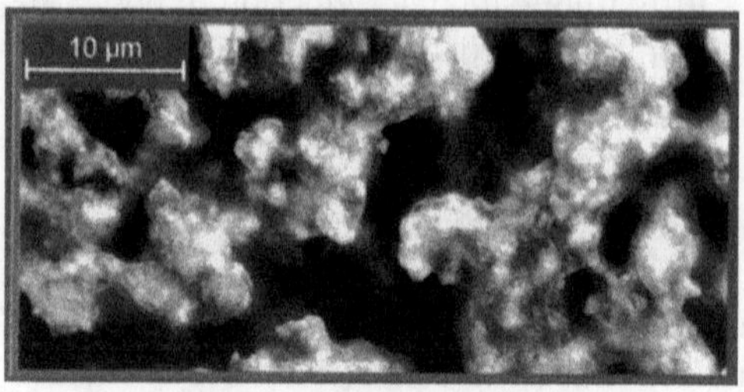

Figure 9.4: Transverse sectional view of structures formed in a powder coating film during curing with a matt hardener (see text for formula); from [118]

Figure 9.5: Scanning electron force micrograph of the surface structure formed during curing with a matt hardener (see text for formula); from [123]

10 Film formation by polymerisation

The bulk of film formation by polymerisation in the coatings sector is effected by **radiation curing** and 2-pack chemical curing of liquid 100 % systems, with the former much more prevalent than the latter. Polymerisation is generally a rapid curing technique.

10.1 General polymerisation mechanisms

10.1.1 Free-radical polymerisation

In free-radical polymerisation, the monomers are molecules which have one or more double bonds between carbon atoms. In paint curing, the role of polyfunctional or crosslinkable monomers is played by resins that contain double bonds, possibly in combination with reactive diluents or dispersed in water .

A double bond does not consist of two equal bonds, but rather of a stable **σ-bond** and an additional, weaker **π-bond**. The latter can be cleaved by the attack of a sufficiently reactive free-radical R• to produce a new free-radical, which in turn attacks a π-bond, etc.:

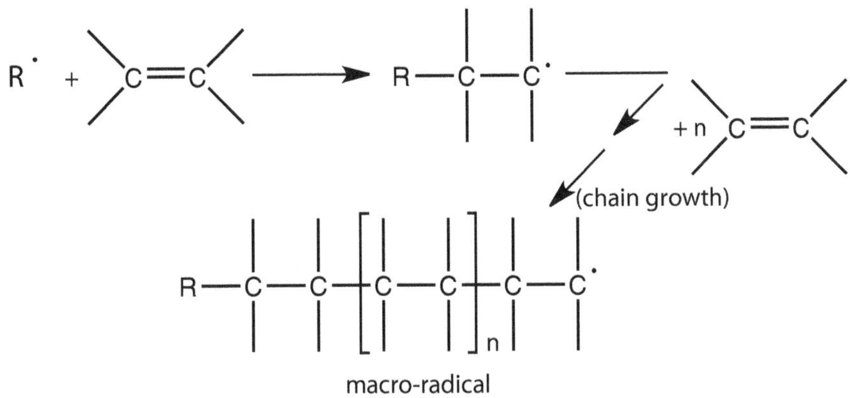

The result - in kinetic terms - is a **chain reaction** that leads to a free-radical molecular chain (macro-radical), which keeps reacting until its free-radical end collides with some other free-radical. This free-radicals combine and chain growth terminates, e.g.

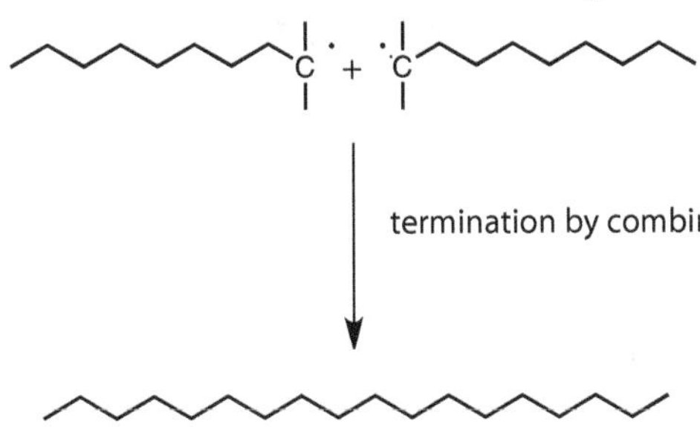

Substituents on the chains are not shown.

Peter Mischke: Film Formation in Modern Paint Systems

ISBN 978-3-86630-861-9

This **termination by combination** leads on average to a doubling of molecular weight, which is beneficial in this case. However, **termination by disproportionation** is also possible. This occurs when an H atom is transferred from one end of a free-radical to another, as a result of which the molecular weight is virtually unchanged. Often, the two termination mechanisms occur in parallel.

The free-radicals which initiate polymerisation are almost always formed by homolysis of an **initiator**, which is incorporated into the polymerising binder as an additive or hardener component. Homolysis can be initiated photochemically, thermally or be chemically accelerated at room temperature:

The first case is the one which obtains during UV curing, while the second and third occur

$$\underset{\text{initiator}}{R{-}R'} \xrightarrow{h\nu \text{ or } \Delta \text{ or catalysator}} R^{\bullet} + R'^{\bullet}$$

with **unsaturated polyesters (UPs)**. It is also possible to bombard the uncured film with a beam of fast electrons at 150 keV or to trigger free-radical formation and subsequent polymerisation without the need for initiator (electron beam curing).

When the end of a free-radical, especially a macro-free-radical, encounters a molecule containing hydrogen (polymer chain, monomer etc), there is a certain, albeit low, probability of **chain transfer**. In chain transfer, the kinetic chain (i.e. the polymerisation process) continues but molecular growth breaks off and a new molecule is formed instead. Particularly noteworthy are transfers involving molecular chains, which lead to branching:

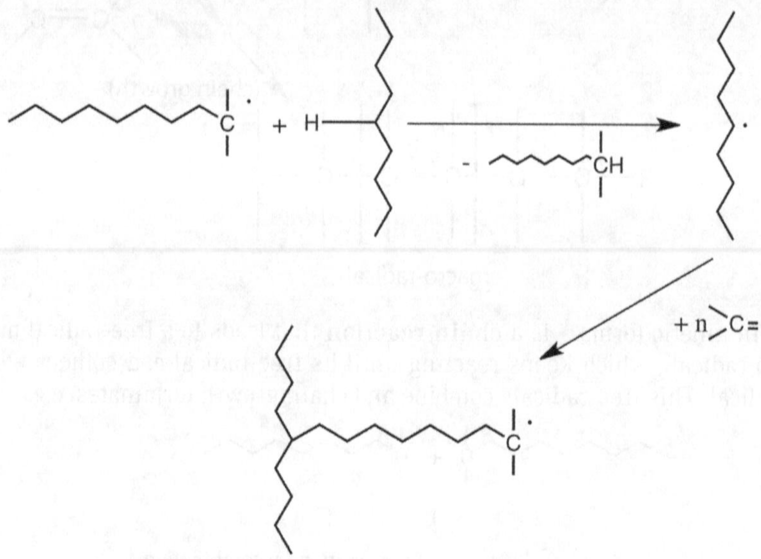

Generally, free-radical polymerisation comes to a halt at once if no more new free-radicals are formed, since reactive free-radicals are very unstable and transient. Free-radicals have the ability to extinguish each other.

10.1.2 Cationic polymerisation

Molecules containing certain heterocyclic rings, such as epoxy groups, or double bonds very rich in π-electrons, such as vinyl (R-O-CH = CH_2), polymerise in the presence of protons formed by the dissociation of strong acids or by photochemical reaction:

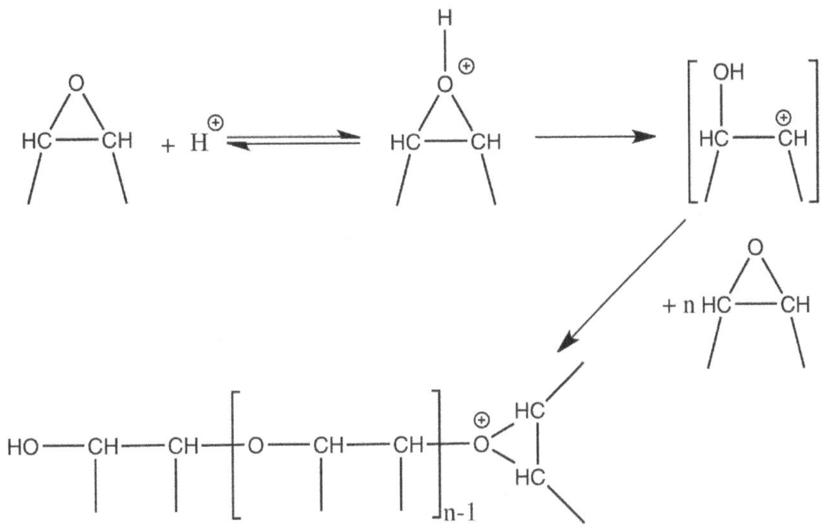

Since the positive chain ends cannot deactivate each other, unlike the case for free-radicals, they are very long-lived, and polymerisation concludes only after all monomer molecules have been incorporated or after **freezing**, unless chain-terminating impurities such as foreign ions or bases are present. Freezing occurs when the polymerisation causes the glass transition temperature to exceed the reaction temperature. The monomers are then unable to diffuse to the reactive chain ends. Typical termination reactions and chain transfer can be formulated as follows:

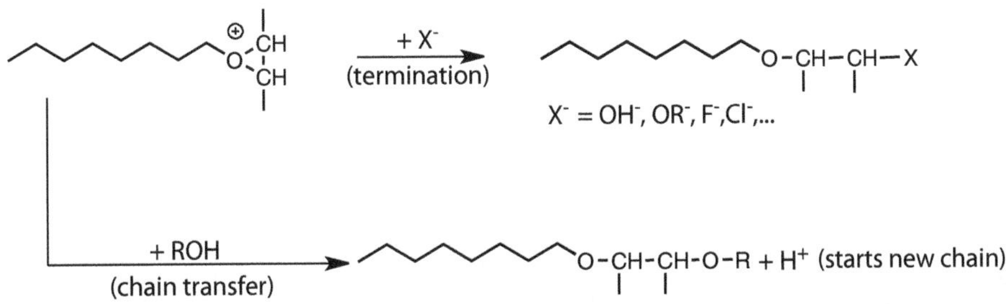

In coating technology, cationic polymerisation is almost only takes the form of ionic curing (see Chapter 10.2.3). Epoxy resins or, more generally, substances with one or more epoxy groups in the molecule can undergo purely chemical cationic oligomerisation or polymerisation with Lewis acids or strong protic acids. One initiator ("hardener") used is **boron trifluoride** blocked with amine [7]. Anionic oligomerisation is possible, too, with alkaline substances, such as tert.-amines, alcoholates and imidazoles [19]. Varying degrees of acid- or base-initiated oligomerisations usually occur at the same time as conventional addition crosslinking of epoxy resins (see Chapter 8.4.2).

10.2 Radiation curing

Radiation curing can be of the free-radical or ionic type. Since the mechanistic principles of the two versions differ significantly, they are treated separately below.

10.2.1 Free-radical curing with UV radiation

10.2.1.1 Initiation

Several physical principles underlying UV radiation were discussed in Chapter 4.3. As mentioned there, UV radiation consists of photons (light quanta) of energy $E = h\,\nu = hc/\lambda$ (h is Planck's constant, ν is the frequency, c is the speed of light, and λ is the wavelength). If the photon does not have enough energy to promote an electron from a **molecular orbital**[22] to an unoccupied molecular orbital of higher energy, there is a certain probability that the photon will be absorbed and the electron promoted. The molecule is then in an excited state, and can react in different ways (see Figure 10.1). The unexcited, ground state of the molecule is symbolised by S_0. S stands for the **singlet state** in which all electrons are paired with opposite spin, and the overall spin is zero. If an electron of the same spin is promoted from a bonding π-orbital or an antibonding n-orbital (free electron pair) to an antibonding π^* orbital by way of first excited state, the molecule finds itself in the S_1 state, which means "first excited singlet state". This is the excited state of lowest energy. However, photons of sufficient energy can occupy higher-energy states (S_2, S_3 etc.), but these immediately revert to the S_1 state, with release of energy. If the excited electron in the S_1 state changes its spin (a process known as **intersystem crossing**), the overall spin is 1 because the remaining electron of the original electron pair is in the ground state. As the overall spin can have three orientations in a magnetic field (+1, 0, -1), this state is called the **triplet state**, T. The various excitation and subsequent processes of a molecule are shown in a **Jablonski diagram**. Figure 10.1 shows a simplified Jablonski diagram, without the vibration sublevel.

Since the S_1 state has a very short lifetime of just 10^{-7} s, the chemical reactions which occur from it direct must be very fast. The T state lasts much longer (10^{-4} s), since transition to the ground state can only occur by intersystem crossing [124, 125].

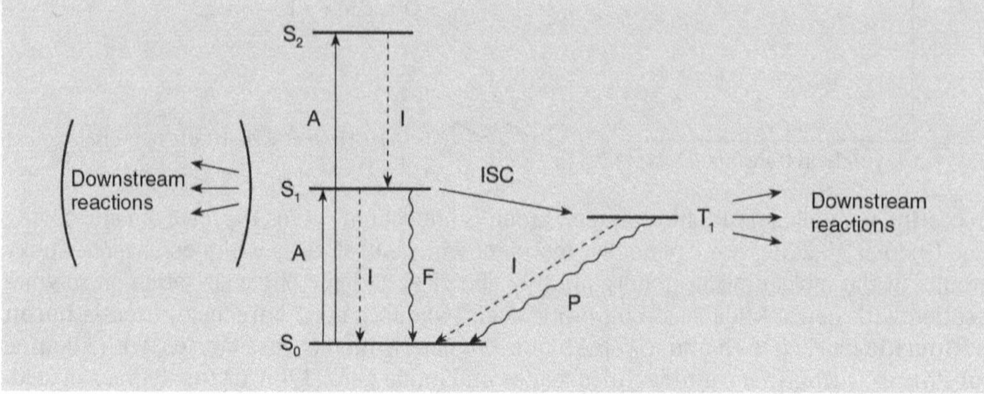

Figure 10.1: The most important photochemical excitation and subsequent processes undergone by a molecule. A is absorption, I is internal conversion, F is fluorescence, P is phosphorescence, and ISC is intersystem crossing

[22] Energy level that extends like a band above the molecule and can be occupied by at most two electrons of opposite spin.

As the excitation energy needed for polymerising double bonds lies in the vacuum UV region, direct photochemical initiation - with rare exceptions (see Chapter 4.3.2) - is not commonplace. It is more usual to take the route involving a **photo-initiator**; this is an additive which absorbs strongly in the UV-B and C regions, but also absorbs photons at longer wavelengths of 300 to 440 nm (the exact wavelength depending on the type of absorber) to form a high **quantum yield** of free-radicals that trigger polymerisation By quantum yield is meant the quotient

$$\Phi = \frac{\text{Number of chemical reaction stages}}{\text{Number of absorbed photons}}$$

Generally, the quantum yield should be as high as possible. For α-cleavage of hydroxyacetophenones, it is 20 to 30 % [126].

With regard to the photochemical mechanism, there are essentially two different types of initiator, α-cleavers and **hydrogen-abstractor-donor systems**. Upon irradiation, α-cleavers decompose monomolecularly by a **Norrish Type 1 reaction** into free-radicals of various reactivities via a triplet state:

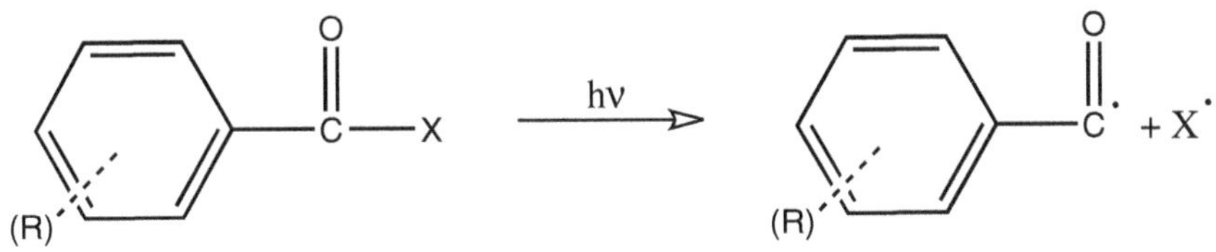

X is a group of widely varying composition that - possibly along with a substituent on the benzene ring - determines the photochemical properties of the photo-initiator. α-Cleavers can be divided chemically into **benzoin ethers**, **benzil ketals**, **hydroxyacetophenones** (hydroxyalkyl phenones) and **acyl phosphinoxides** (APOs). Figure 10.2 shows the long-wave absorption bands of for representatives from these four categories, alongside the transmission band for titanium dioxide.

As may be seen from Figure 10.2, an APO photo-initiator is ideal for curing pigmented paints, provided that the lamp emits wavelengths of the corresponding length. The more recent **bisacylphosphinoxides (BAPOs)** are even more effective than the conventional **monoacylphosphinoxides (MAPOs)** [127, 128]. The lamps used here are of the gallium-doped mercury-silver type (see Chapter 4.3.2).

The other aforementioned type of photo-initiator is a combination of a **hydrogen abstractor** and a **hydrogen donor**. The abstractors include **benzophenones** and derivative classes of compounds, such as **xanthones** and **thioxanthones**. The latter are useful for pigmented systems because of their good absorption of long wavelengths. The hydrogen donors (co-initiators, **synergists**) are usually amines bearing hydrogen atoms on the C atoms adjacent to the N atom. Examples are N-methyl diethanolamine, triethanolamine, methyl-(p-dimethylamino) benzoate and **aminoacrylates**. The last-mentioned of these have the advantage of polymerising into the film during curing, as a result of which odour formation (low amines smell strongly of fish!) and migration are suppressed [5].

Irradiation induces a relatively long-lived triplet state in the hydrogen abstractors. Upon colliding in this excited form with the hydrogen donor, they form an excited charge transfer complex or **exciplex** which reacts further by proton transfer to form two free-radicals [5]:

Of the resultant free-radicals, only the amine free-radical can act as an initiator [22].

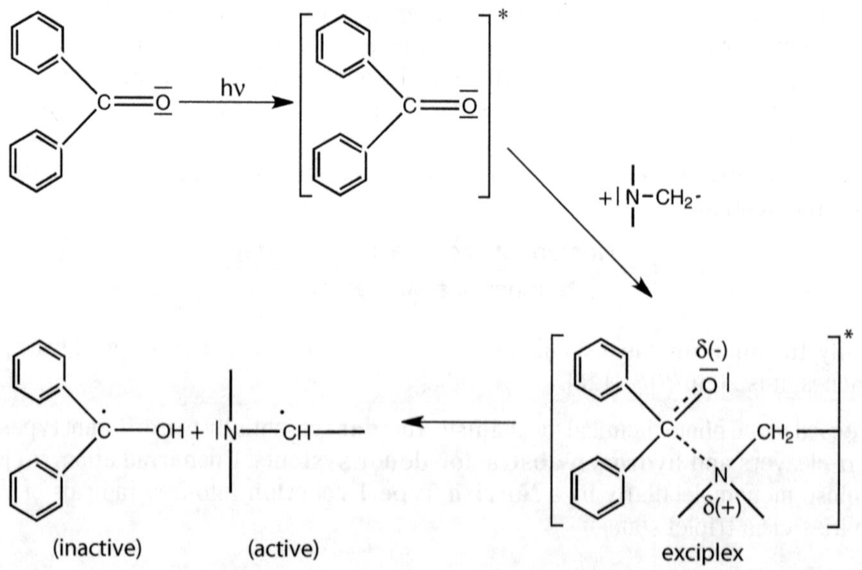

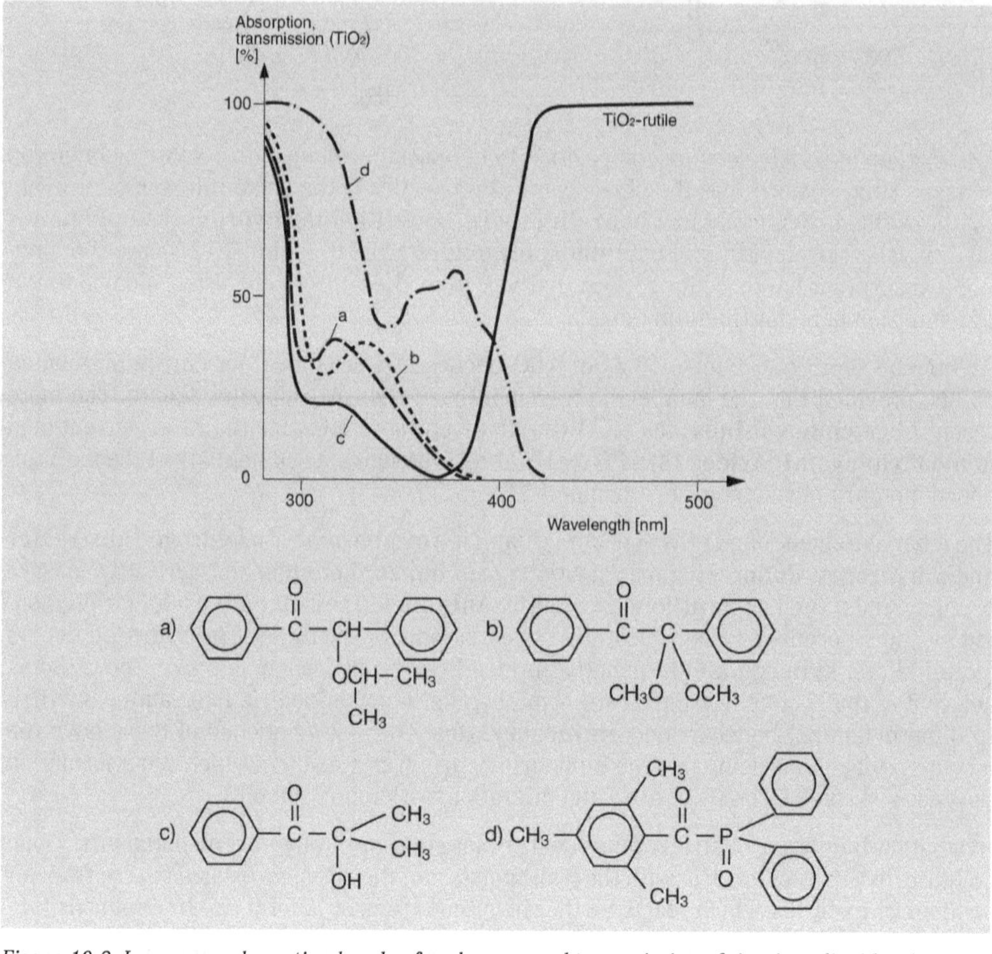

Figure 10.2: Long-wave absorption bands of α-cleavers and transmission of titanium dioxide. a) benzoin ether b) benzil ketal c) hydroxyacetophenone d) acylphosphinoxide; from [7]

The hydrogen abstractor-donor systems exhibit superior surface curing, but poorer deep curing than the α-cleavers. The former property used to be attributed to the presence of the amines, which weaken **oxygen inhibition** on the surface during curing (see Chapter 10.2.1.3) [5], but more recent knowledge suggests that this explanation is inadequate [129]. When UP resins in styrene are cured, the benzophenone systems have hardly any initiating effect. The reason is that styrene absorbs the triplet-excitation energy from the benzophenone (or derivative), without itself forming free-radicals; i.e. the styrene functions as a **triplet quencher** [19]. In practice, a mixture of different photo-initiators is often employed in order that better use may be made of the lamp spectrum and, at the same time, surface and deep curing may be promoted.

It should not go unmentioned here that not only are there α-cleavers, but there are also **β-cleavers** , e.g. α-chloroketones. A further way to effect initiation is to cure maleimides, acrylates or fumarates (electron acceptors) with styrene, allyl and vinyl ethers, and amides or α-olefins (electron donors) by means of alternating photo-copolymerisation. These monomers primarily give rise to charge-transfer complexes, which themselves act as photo-initiators. The XeCl **excimer lamp** (308 nm) is ideal for such systems [12, 22, 128].

Attenuation of the normal incident UV radiation with increase in depth of penetration into the film is given by the **Lambert-Beer law**.

Equation 10.1: $$E(\lambda)=\lg\frac{I_0}{I(x)}=\varepsilon(\lambda)cx$$

$E(\lambda)$ is the molar decadic extinction
I_0 is the incident intensity, I for x = 0 (e.g. in W m^{-2})
$I(x)$ is the residual intensity at depth x
$\varepsilon(\lambda)$ is the molar decadic extinction coefficient
c is the photo-initiator concentration in mol l^{-1}

ε is the concentration- and intensity-independent absorption power of the photo-initiator or absorbing substance generally and thus is purely material specific. If we assume that, for adequate curing, the residual intensity on the substrate under a 50 µm paint layer is 10 % of the initial intensity, then E = 1. For a 250 µm-thick layer, E would equal 5, since x has quintupled while ε and c have remained constant. The relative residual intensity thus computes to 1/100,000, which would probably not be enough for curing the substrate. However, if we make the not unreasonable assumption that E = 1 is kept constant for deep curing, the initiator concentration must theoretically be reduced to 20 % of the original concentration.

The rule is therefore:
- the thicker the layer to be cured, the less photo-initiator needs to be employed

Typical concentrations are 2 to 4 % for thin layers, and less than 1 % for thick layers, equivalent to a molar concentration of 0.05 to 0.2 mol l^{-1}.

Any consideration of radiation absorption must not overlook the dependence on wavelength. Thus, long-wave UV-A radiation (in the case of clear coats) is absorbed almost only by the photo-initiator and, in accordance with the calculation above, can penetrate deeply. In contrast, the short-wave radiation is swallowed up by the binder in the surface and thus contributes to surface curing only [9].

10.2.1.2 Free-radical polymerisation

Viscous **acrylate oligomers** are the most important type of binder for free-radical radiation curing. These are characterised by having several polymerisable **acryloyl groups** attached to a polyester, epoxy resin, polyurethane or polyether backbone:

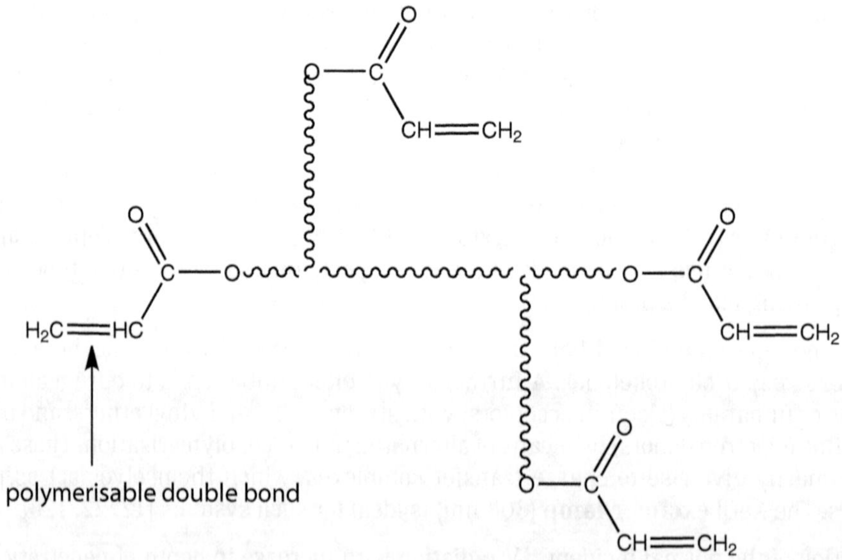

The corresponding resins are polyester acrylates, epoxy acrylates, urethane acrylate acrylates and polyether acrylates. For specialty applications, other base substances may be used, e.g. silicones for release coatings. These liquid resins are either inherently of low-enough viscosity for, e.g. roller application; if they are not, lower-viscosity, low-molecular monofunctional or higher-functional acrylates are added to serve as reactive diluents. Apart from acrylates, UP systems are also employed (see Chapter 10.3).

If the binder additionally contains a photo-initiator, strong UV irradiation leads to very rapid, crosslinking polymerisation; curing generally takes around one second or less. Figure 10.3 shows a schematic diagram of photo-crosslinking.

The glass transition temperature (T_g) rises during curing. If it reaches the instantaneous curing temperature, which can (briefly) be 80 °C owing to the **heat of polymerisation** and the IR content of the lamps, the polymerisation reaction freezes due to the immobility of the monomers or double bonds. If T_g is approx. 20 K higher than the curing temperature, further conversion is virtually impossible – even upon renewed irradiation. Attainable curing conversion (double-bond conversion) is typically between 60 and 90 % and increases with increase in curing temperature [22, 130].

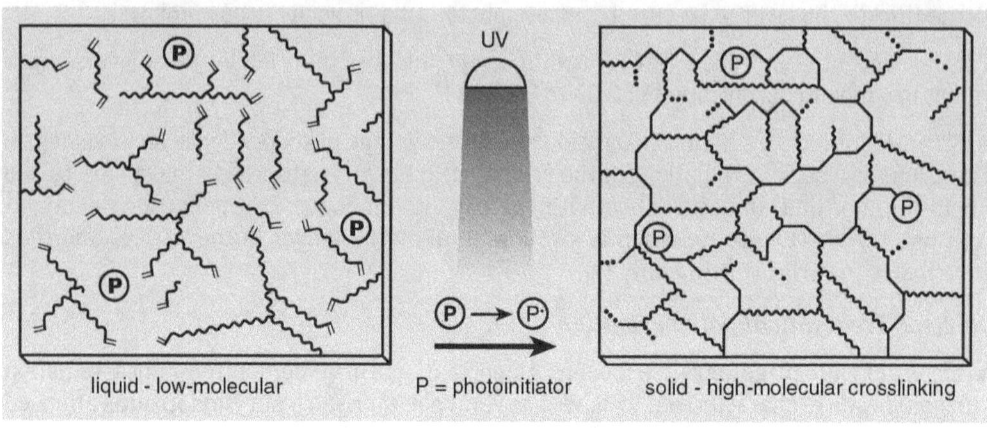

Figure 10.3: Photo-crosslinking (radiation curing) of an acrylic-UV paint; from [9]

Polymerisation of pure monoacrylates should yield linear macromolecules only. In practice, however, minor crosslinking due to chain transfer is observed [12]. Since even a diacrylate bridging two chains has a crosslinking effect, and since the paint generally contains a predominance of even-higher-functional oligomers, polymerisation immediately leads to dense crosslinking. The network formed is highly flawed, though; i.e. it contains numerous free chain ends, possibly with unreacted double bonds (see Figure 10.3). This is reflected *inter alia* by the fact that the **gel point**, i.e. the polymerisation conversion, at which actual crosslinking starts, is 10 to 100 times as high as the calculated value [12].

Meischner et al. used mathematical simulations to study crosslinking in model systems, and compared the predictions with the experimental data [131]. Their results may be summarised as follows:

- the higher the functionality of the paint, the harder the films (pendulum damping)
- the lower the functionality, the higher the conversion; e.g. curing of monoacrylate only: 98 %; curing of triacrylate only: 45 %
- the glass transition temperature (T_g) increases with increase in functionality, and the glass transition (decrease in storage modulus E' on heating) becomes less and less pronounced
- the crosslink density as calculated from E' (experimental value) using Equation 5.4 does not agree with the purely theoretical value; instead, a relationship of the type $\ln E' = m \nu + b$ (ν is the theoretical crosslink density; m and b are constants) is found
- with regard to the last point, the authors point out that Equation 5.4 holds only for ideal networks, which do not exist in radiation-cured paint films.

As explained above, the radiant intensity in the paint film decreases in accordance with the Lambert-Beer law (Equation 10.1). It might be asked why that is so important since, after all, polymerisation continues very rapidly in the film, i.e. at increasing depths. But that would be to misunderstand free-radical polymerisation. If that were the case, the decrease in film hardness from the surface towards the substrate would not be observed as frequently as it is [132]. The reach of growing macro-free-radicals and the diffusion distance travelled by initiator free-radicals formed are very short because they have a short lifetime. Consequently:

- polymerisation or curing occurs only when sufficient initiator <u>and</u> radiation are present!

The long-distance action in fact extends only a few nanometres [22]. Photo-initiator, photo-initiator concentration and radiation parameters (spectrum, intensity and length of irradiation) need to be chosen judiciously if the radiation is to reach all regions of the film. This is particularly the case for pigmented coatings containing high concentrations of photo-initiator.

Excessive absorption of the radiation in the uppermost film layer will cause the surface to cure extensively first and then to shrink, while the lower depths of the film remain soft. Further exposure then causes the lower depths to cure, thereby giving rise to **wrinkling.** This effect is counter-acted, however, by oxygen inhibition (see Chapter 10.2.1.3), which is why it tends to occur more in inert-gas curing or cationic curing [22].

Polymerisation is associated with a **volume shrinkage** of 5 to 10 %, which leads to abherend shear stress on flat, unyielding substrates, such as metal. For this reason, direct UV coating of metals with free-radical-curing, conventional systems is generally deemed problematic.

10.2.1.3 Oxygen inhibition

All free-radical-curing coatings suffer from the basic problem of oxygen inhibition, in which atmospheric oxygen extensively suppresses polymerisation on the surface. Oxygen (O_2) is a fairly stable diradical (see Chapter 5.5.5) that bonds spontaneously to other free-radicals because hardly any activation energy is required:

$$R^\bullet + {}^\bullet OO^\bullet \longrightarrow R\text{-}OO^\bullet$$

The resultant peroxy free-radicals are relatively stable or unreactive. They therefore do not trigger polymerisation and, at best, saturate themselves to become hydroperoxides. Atmospheric oxygen is even used deliberately as an inhibitor for the UV-resin synthesis, where it prevents premature polymerisation in the reactor. Furthermore, oxygen can quench photochemical excitation of the initiator upon contact, before the latter has a chance to become a free-radical [133].

In unfavourable circumstances (no inert gas, weak lamps, too little photo-initiator, no amine addition), inhibition can prevent the film from curing down to a depth of about 10 μm, and so the film remains tacky. Inhibition can usually be countered by increasing the amount of free-radicals in the surface. Polymerisation then "overtakes" post-diffusion of oxygen into the film.

Inhibition by oxygen can be greatly suppressed by adding amines, which also serve as synergists in photo-initiation (see Chapter 10.2.1.1). The effect of an amine can be written as follows [134]:

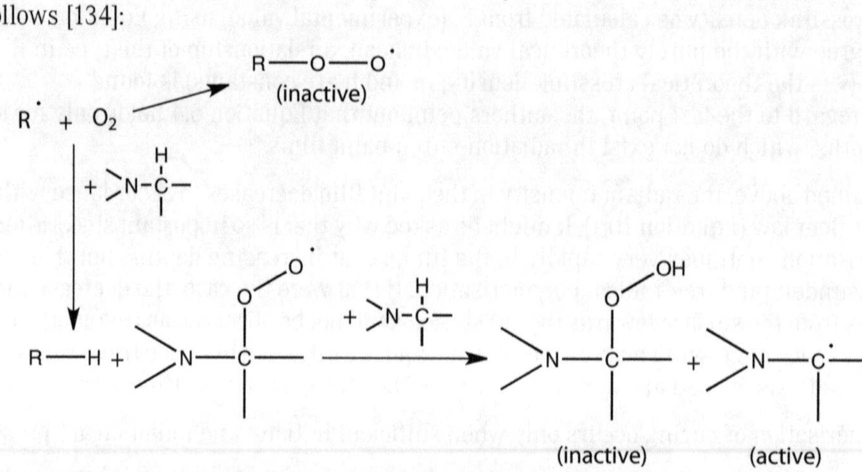

The amine may be bound to the binder, too, to suppress its odour.

Analysis by methods such as confocal **Raman spectroscopy (Raman microscopy)** show that the added amine causes the double bond conversion on the surface to attain virtually the same value as inside the film [135].

While access of air is traditionally prevented – where necessary – with nitrogen, *Beck* has established the advantages of inhibition with carbon dioxide (CO_2). Under CO_2, even weak lamps are capable of effecting tack-free curing of three-dimensional objects, without added amine and with reduced photo-initiator content [133].

10.2.2 Electron beam curing (EBC)

Free-radical curing can be triggered direct by an **electron beam** instead of the combination of UV/photo-initiator. Such a beam consists of countless, essentially parallel "airborne" electrons with a kinetic energy of around 150 keV[23] or more that are emitted two-dimensionally from electron accelerators.

[23] 1 Kilo electron volt; 1 eV is the product of the elementary charge ($1.6022 \cdot 10^{-19}$ C) and 1 V.

The gigantic excess of energy involved becomes apparent when we compare this particle energy with the few eV that constitute the bond energy of a chemical bond or the energy of UV photons. When the electrons (e) collide with the coating molecules, numerous random processes occur which – apart from x-ray emission (see below) – all yield free-radicals. The most important processes are[24]:

- collision ionisation of a molecule AB
  $A{-}B + e_{primary} \longrightarrow A^{\bullet} + B^{+} + 2\,e$
- capture of a slow, ejected secondary electron
  $B^{+} + e_{secondary} \longrightarrow B^{\bullet}$
- Excitation, with homolysis
  $A{-}B + e_{secondary,fast} \longrightarrow A^{\bullet} + B^{\bullet} + e_{secondary,slow}$

Furthermore, various interactions between the electrons and the material give rise to **X-rays**. For health reasons, this is a serious disadvantage of EBC over UV technology.

An electron beam will pass through thick, pigmented coating layers and even penetrate deep into the substrate, where it will fully cure any binder which has penetrated thus far.

The countless free-radicals formed inside the paint layer trigger crosslinking polymerisation – in the absence of a photo-initiator – which is up to ten times as fast as UV curing. The rapid viscosity rise and the immobilising effect of the inchoate network prevents growing free-radicals from colliding with each other, with the result that the number of growing chains and thus the polymerisation rate increases at an explosive rate (**Trommsdorff** or **gel effect**). Unlike the case for UV curing, which proceeds from the surface, EBC requires the use of nitrogen to render the coating surface inert.

Despite the fact that EBC offers certain advantages over the UV process, such as no photo-initiator, much faster curing, greater surface throughput, and coatings of greater durability, it has not gained widespread acceptance.

10.2.3 Ionic UV curing

10.2.3.1 Initiation

The principle underlying cationic polymerisation was presented in Chapter 10.1.2. As mentioned there, it is triggered by protons. In ionic UV curing, proton formation is triggered by photolysis of photo-initiators, such as **sulphonium** (1) and **iodonium** salts (2).

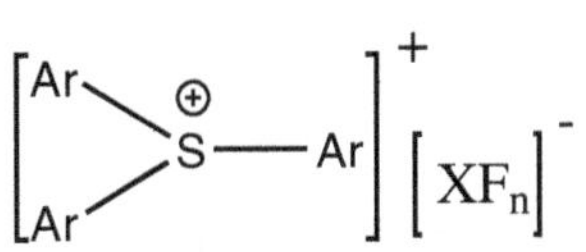

1

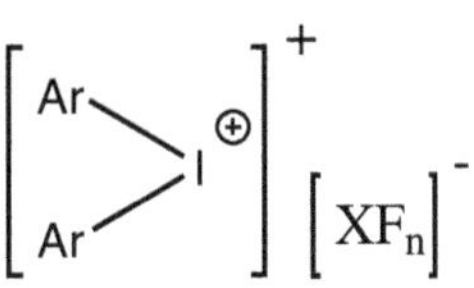

2

Ar represents phenyl groups (including substituted ones)

[24] Not formulated consistently in the literature

$[XF_n]^-$ is $[SbF_6]^-$, $[AsF_6]^-$, $[PF_6]^-$, $[BF_4]^-$; and more recently, also,

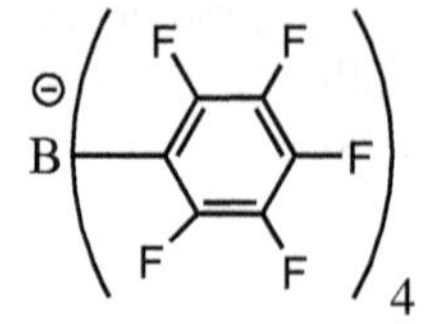

Proton formation proceeds essentially as follows:

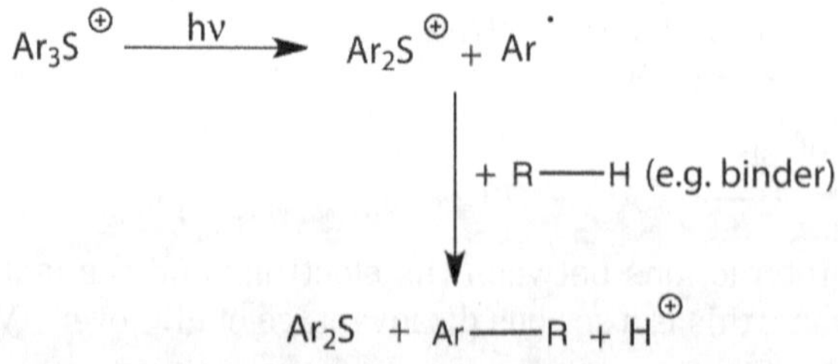

The absorbed wavelength and the absorption intensity are determined by the structure or the substituents of the aryl groups. The strength of the initiation in the form of the curing rate of epoxies, in contrast, depends on the complex anion; it decreases in the order given above [5].

A further class of substance is that of **ferrocenium salts**. These typically have the following structure [5, 17]:

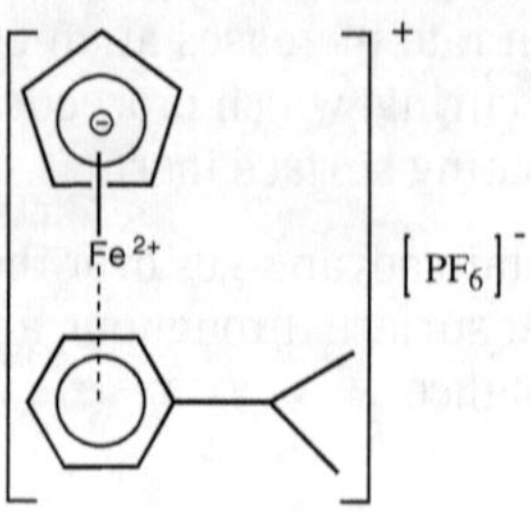

Irradiation leads to cleavage of the cumene ligand. The result is three electron pair holes in the iron atom, which are filled by three epoxy rings, with subsequent ring opening; the latter process requires thermal activation after the irradiation step [126]. Absorption occurs to a depth of roughly 500 nm, which permits very thick layers to be cured in one pass [17].

Onium salts absorb very weakly above 350 nm, and this is disadvantageous for pigmented systems, such as printing inks, which are a primary application area of UV curing. The problems can be eliminated or alleviated by adding **sensitizers** [22]. These are basically the same substances as the hydrogen abstractors used in free-radical curing, where they are also occasionally called sensitizers. Two typical examples are isopropyl thioxanthone (ITX) and 1-chloro-4-propoxy thioxanthone (CPTX):

These absorb in the range 370 to 410 nm and can thus improve the curing of pigmented systems [136]. The principle behind sensitization is shown in Figure 10.4 (see also Figure 10.1).

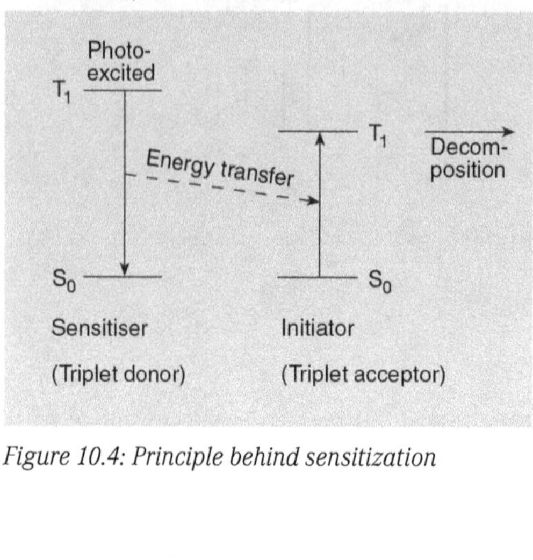

Figure 10.4: Principle behind sensitization

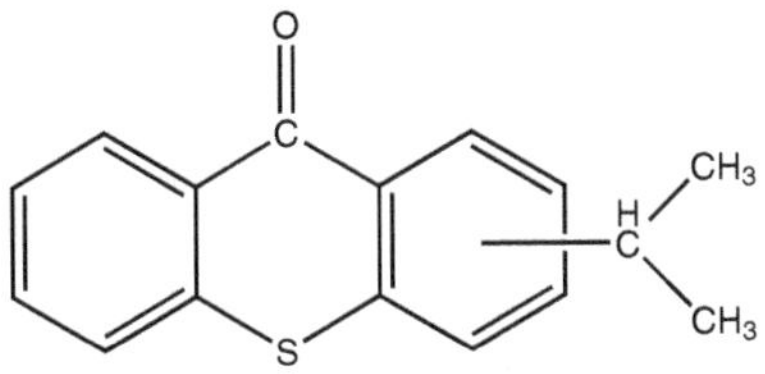

ITX

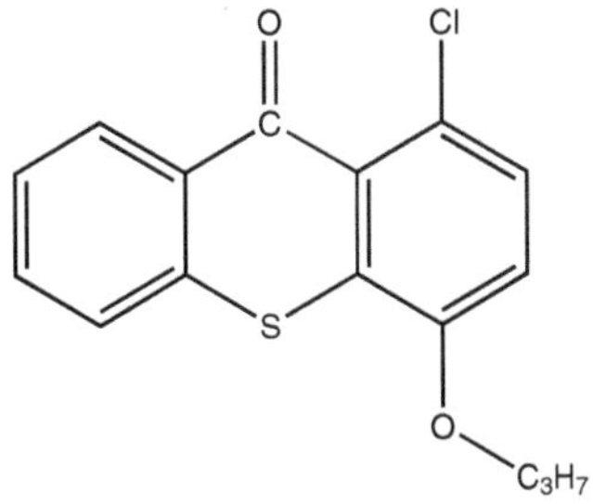

CPTX

10.2.3.2 Polymerisation

The binders employed for cationic UV curing are mainly epoxy resins, which cure by means of ring-opening polymerisation (see Chapter 10.1.2). The conventional bisphenol-epichlorohydrin and other glycidyl-terminated liquid resins show less reactivity, though; this weakness can be compensated for by raising the temperature to 70 to 80 °C, e.g. by additional IR irradiation [22]. Much more reactive are the more common cycloolefin epoxy resins, e.g. 3,4-epoxycyclohexylmethyl-3',4'-epoxycyclohexane carboxylate (ECC):

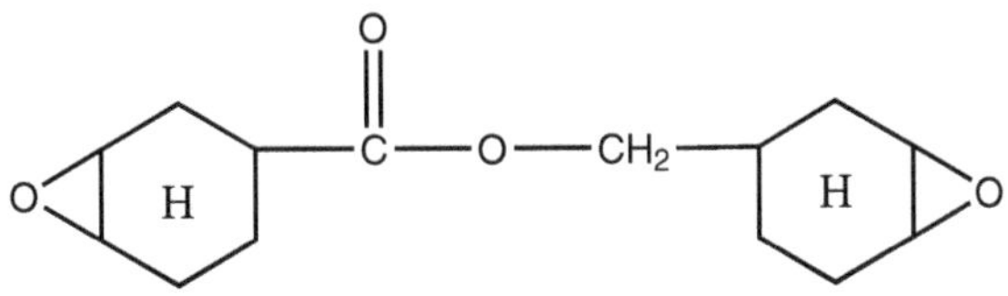

ECC

The greater reactivity of the cycloolefinic resins, relative to the glycidyl compounds, stems on one hand from the greater tension in the oxirane ring and on the other from the fact that the secondary carbocations formed by ring opening are more stable than primary ones on account of hyperconjugation:

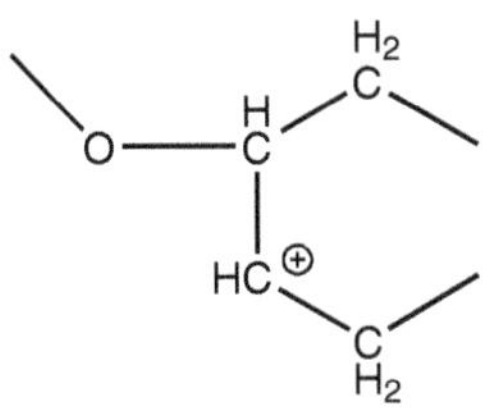

(more stable)

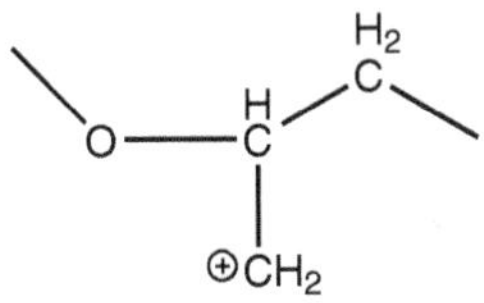

(less stable)

This means that the activation energy for cationic ring opening is smaller and so the latter proceeds faster.

Vinyl ethers, which copolymerise well with the EP resins, can also serve as **reactive diluents** and to further boost the reactivity [5, 22].

The polymerisation or curing rate increases with increase in availability of protons. This may be achieved by raising the photo-initiator concentration and/or the radiant intensity. The optimum level needs to be found for both of these approaches. Conversely, all basic (alkaline) substances in the paint, such as alkaline pigments, have a deleterious effect on curing [5, 12].

Water or alcohols in the paint give rise to **chain transfer** which lowers the resultant molecular weights or crosslink density (see Chapter 10.1.2). But they chiefly lessen the

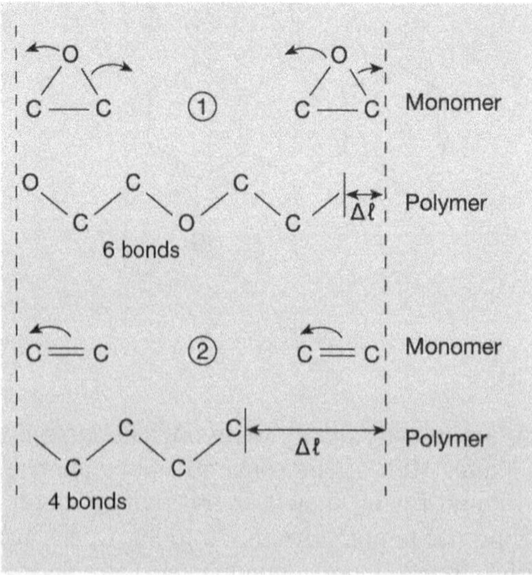

Figure 10.5: Volume shrinkage by polymerisation (not to scale). 1) EP groups (cationic), 2) double bonds (free-radical

proton activity because they can bond protons (an example of the levelling effect of water). The influence of atmospheric moisture on curing is so slight that curing can generally take place under non-dried air [12]; it may even make sense for alcohols (preferably multi-functional) to be present: as T_g is approached more and more, the curing reaction leads to immobilisation of the growing chains and monomers in the film, as a result of which further curing comes to a halt. The lower-molecular alcohol molecules now act to a certain extent as transport agents for cations, by generating new, highly mobile protons from cationic (immobile) chain ends. The overall outcome is better deep curing and flexibilisation of the film [47].

While ionic curing is slower than the free-radical type in light, it spontaneously undergoes **post cure** (dark cure) which can last for more than 24 hours. A limited amount of dark polymerisation due to proton migration also continues in flat areas that received little or no radiation. Dark curing can be accelerated by heating. Even more favourable is to raise the temperature during irradiation, the effect of this being to boost polymerisation substantially [5, 47].

Clear advantages of cationic curing over free-radical curing are

- no oxygen inhibition; therefore no tacky surfaces
- less volume shrinkage (approx. 3 %, i.e. approx. 1/3rd of the value of free-radical curing; see Figure 10.5), and therefore much better adhesion to metal
- dark cure after irradiation

10.2.4 Waterborne binders

Waterborne products can essentially be divided into

- emulsions of normal acrylic oligomers
- dispersions of higher-molecular acrylates, e.g. **acrylated PUR dispersions**

There are also solutions of certain resins and blends of radiation-curable and conventional dispersions [137].

When free-radical radiation-curable coatings are applied to absorbent substrates, such as wood and derived timber materials, lower-molecular components migrate primarily into the pores, thereby escaping subsequent irradiation. It is therefore advantageous to replace the lower-molecular reactive diluents with water, which, however, must have completely evaporated or totally absorbed by the substrate prior to radiation curing if film problems, such as haze, are to be avoided. For this reason, binder manufacturers also offer acrylate oligomers in the form of aqueous emulsions or self-emulsifying products. Once the water has evaporated, the curing mechanism is essentially no different from that of non-aqueous systems.

Radiation-curable dispersions have the ability to undergo physical drying and photochemical curing. The molecular weights are high to very high (M_w in excess of $2 \cdot 10^5$ g mol^{-1}) and the dispersions are nonetheless of low viscosity, since the molecular weight of dispersions is not directly linked to the viscosity. After brief evaporation, e.g. in a blowing zone at 50 °C, the coating is already physically dry. While still warm, this is followed by UV crosslinking with relatively little photo-initiator, and is accompanied by a further rise in T_g and film durability. Areas of the film which do not receive UV radiation are strong enough to resist moderate stress. Durability can be further increased by additional chemical crosslinking with polyisocyanate. Another strength of these binders over highly crosslinked systems is low polymerisation shrinkage, which leads *inter alia* to better adhesion to metals [138, 139].

10.2.5 Dual cure

Photolysis of ionic photo-initiators also leads to the temporary formation of free-radicals (see Chapter 10.2.3.1). Acrylates can therefore be admixed to ionic curing systems because they also polymerise, forming **interpenetrating networks** with the ionically cured binder. Naturally, an α-cleaver could also be added to the ionic photo-initiator to increase the supply of free-radicals [12].

Dual-cure methods in the strict sense are combinations of radiation curing and pure chemical crosslinking, such as UV/2-pack polyurethane (solvent-based or waterborne), and UV/1-pack melamine. The primary aim of these methods is to achieve paint crosslinking on surfaces which are not irradiated and/or, through UV irradiation prior to final curing, to create a coating which is immediately ready for assembly [2, 140-142].

10.3 Curing of unsaturated polyesters

10.3.1 Polymerisation

Unsaturated polyesters (UP) are linear molecules with a medium molecular weight ranging from 1000 to 3000 g mol^{-1}. Naturally, they do not form useful films by physical drying alone. The UP resin chain contains **double bonds** which are **depleted in π-electrons** and which have been incorporated by co-esterification with **maleic anhydride (MA)** or - in specialty cases - with **fumaric acid**. As a result of cis-trans isomerisation during resin synthesis, some 70 to 95 % of the cis-double bonds of MA will have been converted to the more reactive trans-double bonds (fumaric acid double bonds). On the basis of the 24 to 26 % MA used in the synthesis, the mean functionality (number of double bonds per molecule) computes to between 4 and 6. The resins are generally 65 % solutions in **styrene** as reactive diluent and then fully formulated to paints, stopping compounds or moulding compounds. Other reactive diluents, such as allyl compounds, can be co-used to modify properties; they are not widely employed because they are much more expensive and less reactive than styrene. In special circumstances, non-reactive solvents or even water may be used to lower the viscosity. However, this requires specialty resins since the fumaric and maleic acid double bonds need a copolymerisation partner. Incorporated allyl ether double bonds are ideal for this [143].

Curing consists in copolymerising the styrene with the double bonds of the resin, with volume shrinkage of 4 to 8 % (exact figure depends on the resin type), to yield a very hard film in accordance with the following schematic reaction sequence:

All inhibiting substances, along with added stabilisers, are deactivated initially by the free-radical initiator in what is called the **inhibition phase**. This is followed by the **copolymerisation phase**, which is controlled only by the chemical reactivities, and finally by the **diffusion phase** in which the system becomes more and more vitrified. Heating can further increase final conversion [144].

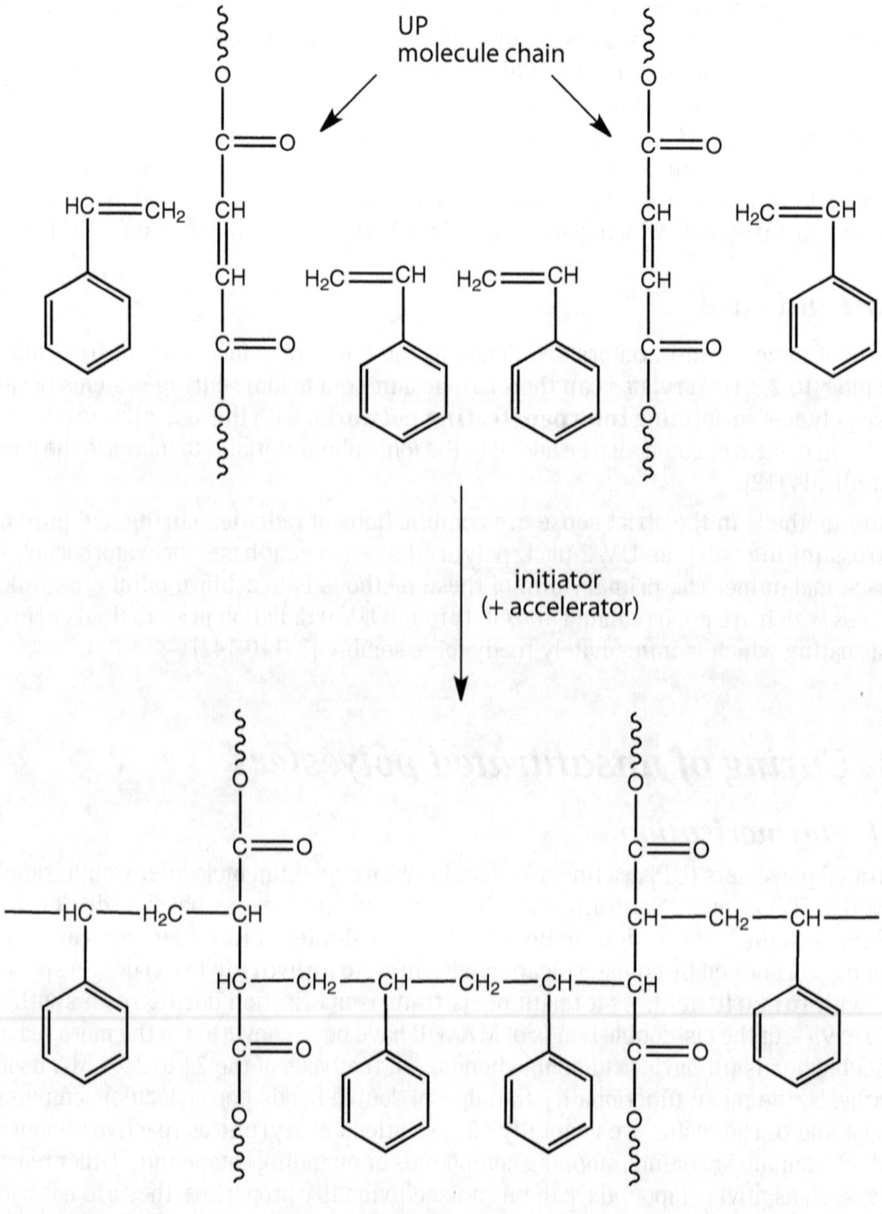

Homopolymerisation of the resin double bonds is not possible. As a quick calculation shows, the number of resin to styrene double bonds is: 1.5 – 2 : 3.5. The bulk of styrene bridges between the chains are thus only 1 to 3 units long. This also fits in well with the fact that styrene prefers to copolymerise in alternating fashion with the resin double bonds. The double bonds of styrene – like the benzene ring itself – are relatively π-electron rich while those of the resin are highly deficient in π-electrons; and as everyone knows: opposites attract[25].

The **copolymerisation diagram** shown in Figure 10.6 for a blend of styrene/fumaric acid diethyl ester shows that, given the double bond fraction of more than 60 % in styrene, the molar fraction of styrene which is incorporated is somewhat less (y_S) than would be expected from its concentration in the starting mixture (x_S). Thus the mixture becomes

[25] Mesomeric stabilisation as per Q-e scheme: $Q_{Styrene} = 1$, $Q_{FAE} = 0{,}75$; Charge: $e_{Styrene} = -0.8$, $e_{FAE} = +1{,}5$ (FAE: fumaric acid ester)

richer and richer in styrene, a fact which, owing to the mobility of the styrene molecules, benefits the final conversion of the fumaric acid or resin double bonds. Maleic acid groups behave much less favourably here; they do not polymerise as readily and so accumulate slowly, as a result of which the residual content of resin double bonds here is higher [144].

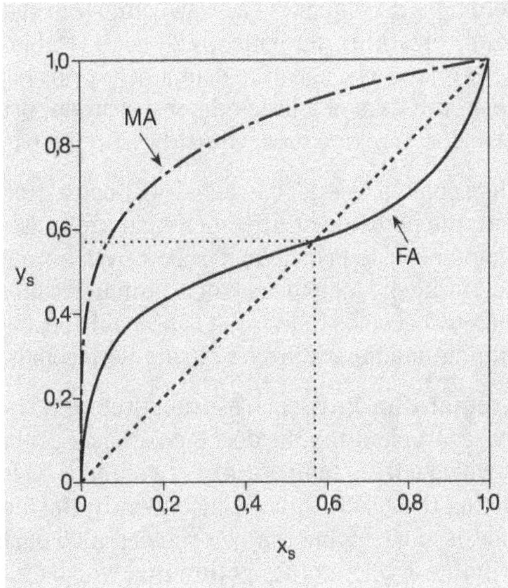

Figure 10.6: Copolymerisation diagram for mixtures of styrene/fumaric acid diethyl ester (FS) and styrene/maleic acid diethyl ester (MS); see text for explanations; from [144]

10.3.2 Initiation, acceleration

Since the boiling point of 160 °C makes styrene very volatile, the paint is generally cured at room temperature. Some heating, e.g. by IR radiation, may be possible. There are two ways to initiate curing:

- addition of photo-initiator (α-cleaver, e.g. benzil dimethyl ketal) and curing under UV radiation
- addition of peroxide initiator and accelerator (activator) and curing without irradiation (2-pack system).

The principle behind radiation curing of styrene is essentially the same as that for acrylates (see Chapter 10.2.1), but it is slower and this ultimately leads to much less surface throughput The UP paints are much cheaper than acrylics, though.

Radiation-curable combinations of UP resins and acrylates have recently been gaining in popularity [145].

Chemical curing can be effected by one of two systems, which can also be combined:

- UP component with **cobalt accelerator** (clear or pigmented/filled) as base paint
  + **Ketone peroxide/**hydroperoxide solution as hardener

or

- UP component with **amine accelerator** as filler/stopping compound
  + **Acyl peroxide** solution as hardener

Two examples of initiators are

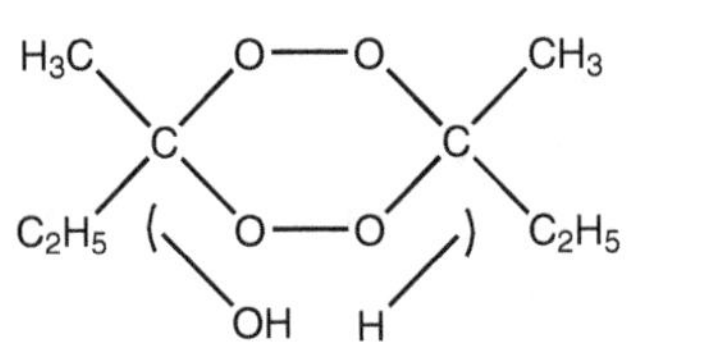

methyl ethyl keton(hydro) peroxide

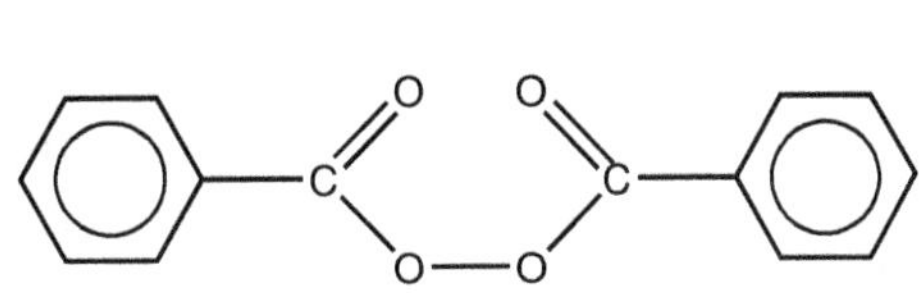

(di)benzoyl peroxide

Peroxides decompose very slowly into free-radicals at room temperature, which is why a UP paint containing peroxide would not be stable in storage. For this reason, the peroxide must be admixed as a separate component prior to application or, e.g., as a liquid active primer , i.e. in the form of a quick-drying solution, prior to coating onto the substrate. In the latter case, the peroxide mixes quickly with the base-paint film by convection and diffusion.

The cobalt accelerator is cobalt (II) octoate or naphthenate. It enters into a redox reaction with the peroxide or hydroperoxide (this was described in the section on oxidative drying; Chapter 7.3). In principle, it is a catalyst, because the Co_2^+ is repeatedly regenerated from the Co_3^+ formed. Since the cobalt accumulates on the pigment and filler surfaces over time and thereby becomes inactive, it is probably best added to the paint just prior to processing. The minimum temperature for curing with cobalt acceleration is quoted as being 15 °C.

Aromatic amines, such as (substituted) N,N-dihydroxyalkyl aniline are particularly effective at accelerating the decomposition of acyl peroxides, such as dibenzoyl peroxide, which is why a curing temperature of around 0 °C is still adequate; however, they are consumed during the acceleration. Consequently, the amount added, namely 0.4 parts to 1 part peroxide, is much higher than for cobalt (0.005 parts Co to 1 part peroxide) [9]. A major drawback of amine accelerators is **yellowing**, which essentially rules out their use in topcoats. On the other hand, the amine undergoes hardly any deactivation in highly filled products. Since handling of amines is really unpleasant for toxicological reasons, manufacturers supply **filler resins** already pre-accelerated, i.e. containing the amines.

Even though the pot lives and curing times of the UP coatings can vary enormously with the resin type, initiator, accelerator, possibly additional accelerator and inhibitor (for storage stability), they are generally rapid-cure systems that normally require 2-pack application methods. They can be **touch-dry** in as little as 15 to 30 min.

10.3.3 Oxygen inhibition

As with free-radical radiation curing of acrylates (see Chapter 10.2.1.3), curing of UP resins suffers from inhibition of polymerisation by atmospheric oxygen at the surface. The surfaces remain tacky to greasy counter-measures are taken. Apart from air-exclusion processing methods, such as **gel coating** and the use of **inert gas**, there are two ways to suppress oxygen inhibition during radiation curing:

- addition of **paraffin wax** (approx. 0.1 %) to the UP paint
- chemical modification of the resin structure

In paraffin addition, the conventional method, paraffin floating on the surface during cross-linking creates a matt protective layer (**paraffin mirror**), which impedes oxygen access and simultaneously reduces styrene evaporation. Owing to the generally unattractive or undesirable matting by the paraffin layer, downstream sanding and polishing are needed.

The second way is to incorporate certain supplementary monomers, such as tetrahydrophthalic acid, endomethylene tetrahydrophthalic acid and allyloxy groups (allyl ether groups). The last of these are particularly effective and are found in **gloss polyesters**. (No matt paraffin layer is formed). The mechanism is that of oxidative post-curing which is initiated by the allyloxy groups and entails copolymerisation of the allyl double bonds with the resin double bonds [1, 19].

Annex 1: The Orchard equation

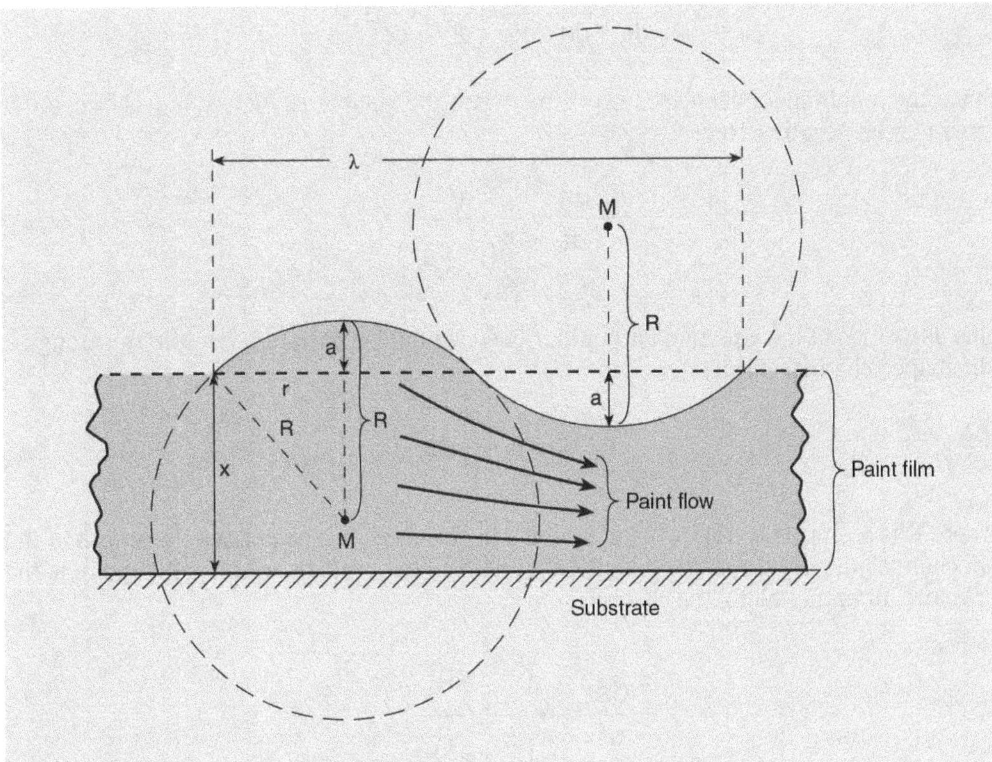

Figure A1: Geometric model for simulating paint levelling

The volume of the spherical section is given by

Equation A1.1:

$$V = \frac{\pi}{3} a^3 (3R - a)$$

The first step is to express R in terms of a and r. From Figure A1,

$$R^2 = r^2 + (R - a)^2$$
$$= r^2 + R^2 - 2Ra + a^2$$
$$0 = r^2 - 2Ra + a^2$$

Equation A1.2:

$$R = \frac{r^2 + a^2}{2a}$$

r is constant during paint levelling and corresponds to one-quarter of the wavelength.

Substituting Equation A1.2 into Equation A1.1:

$$V = \frac{\pi}{3} a^3 \left(3 \frac{r^2 + a^2}{2a} - a \right)$$

$$= \frac{\pi r^2}{2} a + \frac{\pi}{6} a^3$$

Differentiating with respect to time in accordance with the chain rule,

$$\dot{V} = \frac{dV}{dt} = \frac{dV}{da} \cdot \frac{da}{dt} = \left(\frac{\pi}{2} r^2 + \frac{\pi}{2} a^2 \right)$$

Since the amplitude a decreases over time when the volumetric flow V is positive, da/dt must now be negative. Hence:

$$-\frac{da}{dt} = \frac{\dot{V}}{\frac{\pi}{2}\left(r^2 + a^2\right)}$$

One derivation of the equation for laminar flow of a fluid of dynamic viscosity η through a slit-shaped channel of thickness x is

$$\dot{V} = K \frac{x^3}{\eta} \Delta p$$

where K is a proportionality constant, which in this case also comprises the length of the slit channel and the slit width, which are undefined but considered constant, and p is the pressure difference along the channel.

Hence

$$-\frac{da}{dt} = \frac{K \frac{x^3}{\eta} \Delta p}{\frac{\pi}{2}\left(r^2 + a^2\right)\eta}$$

Equation A1.3:

$$-\frac{da}{dt} = K_1 \frac{x^3 \Delta p}{\left(r^2 + a^2\right)\eta}$$

Δp is the difference between the capillary pressures of the same value between a concave and a convex point on the paint film. According to the Young-Laplace equation, the pressure exerted on the liquid by the surface tension σ of a liquid surface of radius of curvature R is

$$p = \frac{2\sigma}{R}$$

where R may also be negative. Thus, the pressure difference is

$$\Delta p = \frac{2\sigma}{R} - \frac{2\sigma}{-R} = \frac{4\sigma}{R}$$

Substituting into Equation A1.3

$$-\frac{da}{dt} = K_1 \frac{x^3 \frac{4\sigma}{R}}{\left(r^2 + a^2\right)\eta}$$

Replacing R by Equation A1.2

$$-\frac{da}{dt} = K_1 \frac{x^3 4\sigma 2a}{(r^2+a^2)\eta(r^2+a^2)}$$

$$= K_2 \frac{x^3\sigma}{\eta} \frac{a}{(r^2+a^2)^2}$$

Resolving the variables

$$-(r^2+a^2)^2 \frac{da}{a} = K_2 \frac{x^3\sigma}{\eta} dt$$

$$-\frac{r^4+2r^2a^2+a^4}{a} da = K_2 \frac{x^3\sigma}{\eta} dt$$

When $a \ll r$, the second and third variables on the numerator on the left side can be ignored and hence eliminated. We then integrate, while allowing for the fact that σ and η can change over time

$$-r^4 \int_{a_0}^{a_t} \frac{da}{a} = K_2 x^3 \int_0^t \frac{\sigma(t)}{\eta(t)} dt$$

$$-\ln \frac{a_t}{a_0} = K_2 \frac{x^3}{r^4} \int_0^t \frac{\sigma(t)}{\eta(t)} dt$$

Finally, substituting $\lambda/4$ for r

Equation A1.4:

$$\ln \frac{a_0}{a_t} = K_3 \frac{x^3}{\lambda^4} \int_0^t \frac{\sigma(t)}{\eta(t)} dt$$

With the exception of the undetermined factor K_3, this equation is the same as the *Orchard* equation (cf. Equation 3.3).

In the literature, $\sigma(t)$ is usually assumed to be constant. Consequently, the fluidity integral appears on the right side.

$$\int_0^t \frac{1}{\eta(t)} dt$$

Annex 2: The WLF equation

Adapted from "Makromolekulare Chemie", Bernd Tieke, 2nd Edition, Wiley-VCH, 2005: From the glass transition temperature T_g upwards, the free volume V_f of a polymer increases roughly linearly with increase in temperature and proportionally to the molecular volume V_M (polymer volume minus the free volume); see Figure 5.7. Thus, we can write:

$$V_f = V_{f,g} + \alpha_f V_M (T - T_g)$$

where $V_{f,g}$ is the "frozen free volume" i.e. the free volume at, or just below, T_g, and α_f is the coefficient of thermal expansion of the free volume. Dividing by V_M:

Equation A2.1:

$$f = f_g + \alpha_f (T - T_g)$$

$$\text{mit } f = \frac{V_f}{V_M} \qquad \text{und } f_g = \frac{V_{f,g}}{V_M}$$

The relationship between the viscosity (η) and V_f is described by the semi-empirical **Doolittle equation**

$$\eta = A e^{B\frac{V_M}{V_f}}$$

Expressed in logarithms

Equation A2.2:

$$\ln\eta = \ln A + B\frac{V_M}{V_f}$$

(A, B are constants)

Formulating Equation A2.2 for the variable temperature T and specifically for T_g, and taking the difference

$$\ln\frac{\eta_T}{\eta_{T_g}} = B\left(\frac{V_M}{V_f} - \frac{V_M}{V_{f,g}}\right)$$

$$= B\left(\frac{1}{\frac{V_f}{V_M}} - \frac{1}{\frac{V_{f,g}}{V_M}}\right)$$

$$= B\left(\frac{1}{f} - \frac{1}{f_g}\right)$$

After substituting Equation A.2.1

$$\ln\frac{\eta_T}{\eta_{T_g}} = B\left(\frac{1}{f_g + \alpha_f (T - T_g)} - \frac{1}{f_g}\right)$$

Combining the fractions

$$\ln\frac{\eta_T}{\eta_{T_g}} = B\frac{-\alpha_f(T-T_g)}{f_g(f_g+\alpha_f(T-T_g))}$$

$$= \frac{-\frac{B}{f_g}(T-T_g)}{\frac{f_g}{\alpha_f}+(T-T_g)}$$

For amorphous polymers, the following values can be inserted to an approximation into the interval T_g to $T_g + 100\,K$

$\alpha_f = 4.8 \cdot 10^{-4}\,K^{-1}$ $f_g = 0.025$ $B = 1$

This yields

$$\ln\frac{\eta_T}{\eta_{T_g}} = \frac{-40(T-T_g)}{52K+(T-T_g)}$$

Values of 40.2 and 51.6 are frequently taken for the constants.

Taking $\eta_{Tg} = 10^{12}$ Pa s as a guide figure for the viscosity at T_g, we arrive at Equation 5.7:

$$\ln\frac{\eta_T}{Pas} = 27{,}6 - \frac{40(T-T_g)}{52K+T-T_g}$$

This equation (WLF equation) also holds in principle for reference temperatures other than T_g. However, the constants will then have different values.

Annex 3:
The gel-point theory of Flory and Stockmayer

Assumption: $N_{A,0}$ $N_{B,0}$ Number of A and B groups at time t = 0
$N_{A,t}$ $N_{B,t}$ Number of free A and B groups still present after time t

Conversion of the A and B groups after time t (from the start of the reaction) is given by

$$U_A = \frac{N_{A,0} - N_{A,t}}{N_{A,0}} \qquad U_B = \frac{N_{B,0} - N_{B,t}}{N_{B,0}}$$

For the monomers with A groups (A monomers), first assume that bi- and tri-functional molecules exist. This can be used to create the following auxiliary variable

$$\gamma = \frac{N_{A,0} \text{ in trifunctional molecules}}{N_{A,0} \text{ total}}$$

The probability P that a group A or B has reacted is

$$P_A = U_A \text{ and } \qquad P_B = U_B$$

Thus, the probability that a branched chain of the following structure will occur is

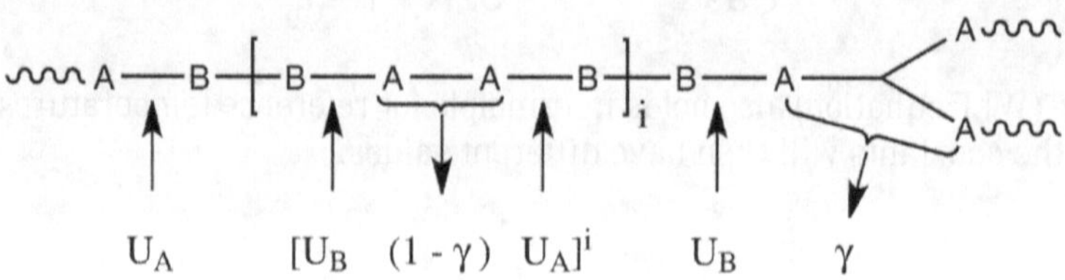

$$W_{ges} = \sum_{i=0}^{\infty} U_A U_B \gamma [U_A U_B (1-\gamma)]^i$$

$$= U_A U_B \gamma \sum_{i=0}^{\infty} U_A U_B (1-\gamma)^i = \alpha$$

This is a geometric progression with factor

$$q = U_A U_B (1-\gamma)$$

Since

$$\sum_{i=0}^{\infty} a q^i = a \frac{1}{1-q}$$

it has the total

Equation A3.1:
$$\alpha = U_A U_B \gamma \frac{1}{1 - U_A U_B (1-\gamma)} = W_{ges}$$

called the branching coefficient. When α = 1 (each chain branches into two further chains)

$$U_A U_B \gamma = 1 - U_A U_B (1-\gamma)$$
$$= 1 - U_A U_B + \gamma U_A U_B$$
$$U_A U_B = 1$$

i.e. every group is reacted: $U_A = U_B = 1$. This is naturally only possible in the case of an equimolar (1:1) preparation. In other words: One chain becomes 2, then 4, then 8 etc. Chain propagation thus follows a geometric progression with factor k = 2. Consequently, since α = 1: k = 2 α.

A geometric progression becomes divergent when k = 1, which would mean

$$k = 2\alpha = 1, \qquad \alpha = 1/2$$

This is interpreted as follows: Every existing chain on average leads to the creation of a new one, and this state corresponds to gelation.

If the branching A monomer were tetrafunctional instead of trifunctional, then

$$3\alpha = 1, \alpha = 1/3$$

since one chain leads to a maximum of three other chains This relationship can generally be expressed by

Equation A3.2:
$$\alpha_G = \frac{1}{f-1}$$

where f is the functionality of the branching monomer and G stands for gelation.
α_G is called the critical branching coefficient.

In Equation A3.1, U_B can be expressed by U_A via the stoichiometric ratio of reactants r

$$r = \frac{N_{A,0}}{N_{B,0}} = \frac{\frac{N_{A,ex}}{U_A}}{\frac{N_{B,ex}}{U_B}} = \frac{U_B}{U_A} \leq 1$$

i.e. $U_B = r\,U_A = r\,U$,

where A is the group which is present in deficiency (and which determines the extent of conversion) and $N_{A,ex} = A_{B,ex}$ is the number of reacted A and B groups. α thus becomes

$$\alpha = rU^2\gamma \frac{1}{1-rU^2(1-\gamma)}$$

and specifically for the start of gelation

$$\alpha_G = rU_G^2\gamma \frac{1}{1-rU_G^2(1-\gamma)}$$

U_G is the conversion at that point, i.e. the gel point. Combining with Equation A3.2 and solving for U_G

Equation A3.3:
$$U_G = \frac{1}{\sqrt{r[1+\gamma(f-2)]}}$$

This equation can be rendered more intelligible by introducing the functionality $\bar{f}_A$ of the A monomers. When only bifunctional and f-functional monomers are present (moles):

$$\begin{aligned}\bar{f}_A &= \frac{n_A(1-\gamma)2+N_A\gamma f}{n_A}\\ &=(1-\gamma)2+\gamma f\\ &=2-2\gamma+\gamma f\\ &=2+\gamma(f-2)\\ \bar{f}_A-1&=1+\gamma(f-2)\end{aligned}$$

Thus, Equation A3.3 can also be written:

Equation A3.4:

$$U_G=\frac{1}{\sqrt{r(\bar{f}_A-1)}}$$

The arithmetic mean computed above for bifunctional and f-functional A molecules can be applied to a mixture of A monomers of arbitrary functionality:

$$\bar{f}_A=\frac{\sum n_{A,i}f_{A,i}}{\sum n_{A,i}}=\sum\left(\frac{n_{A,i}}{\sum n_{A,i}}\right)f_{A,i}$$

It has been observed empirically that Equation A3.4 better describes reality when the weight average of the functionality is used instead of the arithmetic mean:

$$\bar{f}_A=\frac{\sum n_{A,i}f_{A,i}^2}{\sum n_{A,i}f_{A,i}}=\sum\left(\frac{n_{A,i}f_{A,i}}{\sum n_{A,i}f_{A,i}}\right)f_{A,i}$$

The equation for B monomers of higher functionality than 2 in the mixture is (derivation not shown):

$$U_G=\frac{1}{\sqrt{r(\bar{f}_A-1)(\bar{f}_B-1)}}$$

This equation is identical with Equation 8.5.

Annex 4: Test methods used in film formation

Simpler applications testing (relative methods)

A great many properties of a finished coating depend on the degree of drying or curing and may therefore be looked upon as an indirect measure of the degree of film formation. The Pendulum Damping test (ISO 1522) is one such example: this utilises the fact that vibration damping decreases with increase in the degree of drying. Damping depends on several individual parameters:

- ratio of elasticity to viscosity (plasticity)
- ratio of surface drying to deep drying or hardening, and
- tack

The correlation between pendulum damping and degree of drying is therefore complex, and pendulum damping needs to be seen as a semi-quantitative, relative measure of the degree of film formation.

A selection of test methods are provided below that provide direct information about the degree of film formation.

Determination of the degree of drying

Fine glass spheres are scattered over the film. If the spheres can be removed cleanly with a small, soft brush, the degree of drying is rated as 1. A series of loads is then applied for 60 s to a sheet of filter paper placed on the film. In each case, the load is removed and the test panel placed upright. The degree of dryness (on a scale of 2 to 7) is the load at which the filter paper falls off without leaving behind a mark on the film (EN ISO 1517, DIN 53 150).

Degrees of drying include

- set to touch
- dust-free
- dry to touch
- dry-through and
- stackable

Determination of drying time (drying duration, drying period)

The drying time is measured with the aid of a drying recorder. A stylus is pulled at a defined, variable speed over a film which has been knife-coated onto a long, narrow glass panel. The traces left by the stylus are then evaluated. No lasting trace is left behind on a coating exhibiting adequate levelling. Other gradations on the scale are: basic track, film tearing, surface trace and finally no trace (dry). An advantage of this method is that it can be run unattended overnight (ASTM D 5895 A).

Measuring the viscosity increase during drying

An old, yet simple way to measure the change in viscosity during drying is the rolling-ball method of *Wolff* and *Zeidler*. Immediately after the coating material has been applied to a test panel, the panel is inclined at 60° to the horizontal and 2.5-mm steel balls are allowed to roll down it. The speed at which the balls roll down is a measure of the instantaneous viscosity. The readings are substituted into a semi-quantitative equation to calculate the viscosity.

Dissolving, swelling and staining

In general, the more completely dry or fully crosslinked a coating film is, the more slowly or incompletely it dissolves or swells. A cotton ball soaked in a solvent, such as xylene or butanone (methyl ethyl ketone, MEK), is placed on the horizontal coating, and covered with a petri dish. Generally, the longer the exposure, the more the film swells or dissolves; in some cases, it might even lift. Dissolution of the film indicates a high proportion of thermoplastic (sol fraction) while extensive swelling indicates a low crosslink density. Often, both effects occur together. No changes are seen in the case of very densely crosslinked films. A certain degree of softness is often observed when the film is scratched. The results are quoted on a scale of 0 (no change) to 5 (completely destroyed) for the specific exposure period and temperature.

There are several industry-specific standards on how to perform media-resistance tests. For example, EN ISO 2812 ("Determination of resistance to liquids") comprises various tests involving immersion, absorbent media and spotting.

Furniture coatings, in particular, are tested for their resistance to staining by coloured media, such as blackcurrant juice, coffee and dye solutions (EN 12 720). Coatings raw materials are often tested in the opposite manner: the goal here is to remove a fluorescent dye from the coating. The more densely crosslinked the cured film, the harder it is to dissolve the dye.

Determination of blocking resistance or block resistance

Blocking describes the situation which occurs when two coated surfaces which have been pressed together adhere firmly to each other. The cause of blocking is poor surface drying or curing. The films, coated onto test panels, are pressed together under a defined force for a certain length of time. The force necessary to separate the panels is measured and/or the films are inspected for damage (EN ISO 4622).

Wet-scrub resistance of dispersion paints

The more firmly the filler and pigment particles in an emulsion paint film are held together by the binder, the better is its wet-scrub resistance. An abrasive cloth soaked in detergent is rubbed 200 times over a film coated onto a plastic carrier and removal of a test soiling agent and the quantity of film abraded are determined (EN ISO 11 998).

Determining the reactivity of coating powders

Determination of gel time

A small amount (spatula tip) of the coating powder is stirred in a well with a needle, usually at stoving temperature (160 to 200 °C), until threads indicative of incipient gelation start to occur. The elapsed stirring time until gelation occurs is called the gel time. (DIN ISO 8130-6).

Tablet run-off

A tablet pressed from 0.5 g of coating powder is heated on a thin glass plate or metal panel for five minutes at 80 °C to immobilise it. The panel is then placed at an angle of 60° to the horizontal in the laboratory drying oven, which is at stoving temperature. The tablet melts and runs down the panel, leaving a track behind. The length of the track is a qualitative, inverse measure of the curing rate or reactivity. The melt layer is also inspected for gas occlusions, structures, lustre and colour variations.

Scientific methods of determining the degree of cure or conversion

Differential scanning calorimetry (DSC)

Differential scanning calorimetry (DSC) is a micro-calorimetric measuring method that is being used increasingly in applied coatings research. It entails heating two small, sealed metal pans at exactly the same predetermined rate. One pan may contain a small sample of paint or paint film, while the other either may contain an inert reference substance of virtually constant heat capacity or is empty. A specific heat flux is entailed in heating each pan. Initially, i.e. at low temperature, both heat fluxes are virtually constant. However, if the heat capacity of the sample changes - e.g. as may occur at the glass transition temperature (T_g) - or if it undergoes exothermic or endothermic processes, such as chemical reactions or phase transitions (evaporation, melting or crystallization), a significant change occurs in the heat flux into the sample pan that is needed maintain temperature parity between the pans. This difference in heat flow is the primary analyte signal.

In coatings research, DSC is primarily used to determine the T_g, curing temperature and curing conversion (degree of curing). The last of these is determined by heating both an uncured paint sample and a sample of the film formed by the same paint to a temperature higher than the maximum curing temperature. The paint sample yields the total reaction enthalpy of the paint, while the film sample yields the residual reaction enthalpy. The difference between the total and the residual reaction enthalpy is proportional to the curing conversion of the film sample prior to post-curing in the DSC device.

DSC measurements of coatings heated at different rates can also be used to determine kinetic parameters of the curing reaction and sometimes even the reaction mechanism. A complementary variant of this method consists in applying a periodic modulation to the heating rate, affording overlapping signals which can be separated mathematically (MDSC).

Dynamic mechanical analysis (DMA)

In this method, a material sample is set into slight periodic deformation of variable amplitude and frequency. The analyte signal (response signal) is primarily an enforced oscillation, which is converted into the secondary variables of storage modulus (E'or G'), loss modulus (E" or G") and loss factor (tan δ). The sample can be a self-supporting paint film, a dry film placed between two panels or a coating material which has been applied to a support fabric, such as silk, and is still capable of flowing. Measurements can be performed either isothermally at rising frequency or at constant frequency and rising temperature. The second method is particularly interesting for film formation because it allows *in situ* tracking of the change in the material's viscoelastic behaviour from the start of drying until full curing. The storage modulus (above the glass transition range) increases with increase in network density and - in ideal networks - is even directly proportional to the crosslink density. DMA can also be used to accurately determine the glass transition temperature of paint films, and the interaction of storage and loss modulus can be used to draw conclusions about other physical polymer properties.

Infrared (IR) and Raman spectroscopy

These methods are not polymer or paint specific, but they can often deliver direct semi-quantitative information about the concentration of certain functional or atomic groups in coatings and films produced therefrom, without difficult sample preparation. Determining the concentration of reactive groups or their reaction products affords a way of establishing the degree of curing in films. The literature is full of examples of such studies. ATR-IR (attenuated total reflection infrared) spectroscopy yields selective infrared spectra of surface layers, including spectra of films which are impervious to radiation.

Double-bond conversion of acrylate and UP systems is now mostly studied by the related technique of Raman spectroscopy, as it provides stronger signals and clearer spectra. Confocal Raman spectroscopy even permits direct determination of the depth profiles of the double-bond concentration.

Other methods

Examples include measuring the change in a film's dielectric properties (dielectric analysis, DEA) and its reflection of ultrasound waves during curing.

Author

Prof. Dr. **Peter Mischke**, born in 1951, studied chemistry at Hamburg University. After obtaining his doctorate, he went into industry where he worked primarily on metal surface pretreatment and then in the development of synthetic resins. In 1991, he was appointed professor of "Paint Chemistry and Process Engineering" at the University of Applied Sciences in Niederrhein, Krefeld. He teaches chemo-technical subjects as well as polymer, binder, interface and colloid chemistry for paint engineers. He also serves as a lecturer at various vocational training events.

Index

2-pack systems 17
2-pack- = two-component 17
100 % system, liquid 135
α-cleavers 161
β-cleavers 163

A

absorption coefficient 48
absorptions, degree of 48
accelerator 85
  for UP curing 173
acid catalysis 141, 143
acid curing 116
acrylated PUR dispersions 170
acrylate oligomer 163
acrylic powder coating 151
acryloyl group 163
acyl phosphinoxide 161
aggregate structures 60
alkyd resin 20, 105
  long-oil 107
  medium-oil 107
  short-oil 107
allyloxy group 107
amide formation 132
amine 166
aminoacrylate 161
APO = acyl phosphinoide 161

B

back-spraying 31
Baeyer strain 136
barrelling 20
base paint 17
benzoin 155
  ethers 161
benzophenone 161
binder 9
blocked polyisocyanate 133
blocking 101
  agents 133
  strength 66
boron trifluoride 159

C

capillary forces 97-99
carbamate 138
carbonyl dispersion 129, 130
carboxy-epoxy reaction 137, 148, 150
Carothers equation 118
  modified 118
catalysis 116
catalyst 85
cathodic electrocoating 140
CH acid crosslinker 128
chain reaction 117
chain transfer 158, 169
chalking 148
charge-transfer complex 163
CO_2 bubble formation 133
coalescing agent 101
coating material; physically drying 16
coating, inorganic 9
cobalt accelerator 173
coil interpenetration 60
co-initiator 161
combination
  drier 112
  termination by 158
complementary resin 101, 116
complexing agents 113
conversion layers 21
copolymerisation diagram 172
core-sheel morphology 101
cosolvent 76
craters 150
crinkle finish 112
crosslink density 61
crosslinker
  CH acid 128
  resin 116
crosslinking rate 164
curing of UP 173

D

DBTL = dibutyl tin dilaurate 131, 133
deblocking temperature 133
deep curing 163
degree of etherification 141

degree of polymerisation
average 117
dendrimer 119
dewetting 34
dicyanodiamide 149
diffusion coefficient 78
diffusion constant 122
dihydrazide 130
disperse systems, nonaqueous 104
dispersing agents 12
disproportionation, termination by 159
double bond 157
conversion 164
isolated 106
droplet mountain range 36, 111
dual cure 171

E

edge
build-up 28
line 33
pulling 41, 145, 154
electrocoagulation 20
electrodeposition, cathodic 20
electron beam curing (EBC) 54, 26, 166
electroosmosis 21
electrostatic spraying slot 27
elementary reaction 81
elimination-addition mechanism 134
emissivity 47
emitters
long-wavelength 47
mercury vapour lamps 51
emulsifier 76, 103
energy, activation 81
entanglement 60
length 60, 71
epoxy acrylate 164
epoxy amine system, waterborne 138
epoxy group 135
internal 135
epoxy powder coating, curing
with dicyanodiamide 149
with polyanhydride 149
with polyphenol 149
epoxy resin 126, 169
(cyclo)olefine type 135
dispersion 138
emulsion 138
epoxy resin curing, with
amines 137
amino hardeners 138
reaction mechanism 136
via OH groups 140
epoxysilane 129
epoxysilane crosslinker
equilibrium constant 83
equivalent mass 120, 83
excimer lamp 163
exciplex 161

F

Faraday cages 28
ferrocenium salt 168
fibre board, medium (MDF) 30
Fick's law, first 78
filler 11
filler resin 174
filling 9
Flory and Stockmayer, theory of 119
flow behaviour
Newtonian 15
non-Newtonian 15
flow-coating 19
flow equilibria 81
flow improver 155
formaldehyde condensation resin 125
forward roller coating 19
freezing 159

G

gel 144
gelation 118
gel fraction 62, 104, 120
gel point 118, 151, 165
glass transition 62
gloss polyester 174
glycidyl ester 135, 149

H

Hamaker constant 99
hardeners, amino 137
heating time 153
high boilers 67, 27
high solid binder 146
high solid coating 40, 145
hot-spraying 24
hybrid
binder 145
coating 148
powder coating 148
hydrogels 62
hydrogen abstractor 161
donor system 161
hydrogen bonds 59
hydrogen donor 161
hydroxyacetophenone 161

I

induction drying 54
induction period 109
infrared (IR) radiation 45
initial crosslinking temperature 134
initial deblocking temperature 134
initiator 158
in situ adduct 138
internal epoxy groups 135
internal plasticisation 60
interpenetrating networks 171
intersystem crossing 160
iodine value 106
iodonium salt 167
isocyanate 106
  -prepolymer 126
  group 115

J

Jablonski diagram 160

K

ketone peroxide 173
knife coating 28

L

Lambert-Beer law 163
lamp
  arc 51
  electrodeless 52
  excimer 52, 163
  high-pressure 51
  medium-pressure 51
levelling additive 155

M

mass action, law of 82
mass transfer coefficient 87
matt hardener 156
matting 155
MDF 30
metallic paints 48
methylene group 106
methylol group 140, 139
MFT 100
modulus of elasticity 62
molecular orbital (MO) 160

N

NAD = non aqueous dispersion 104
nanocomposites 145
network
  arcs 61
  defects 62
  interpenetrating 171
  mesh 61
  node 61
  structure 92
Nix and Dodge, equation of 153
non-solvent, diluent 92

O

object temperature 123
OEM = original equipment manufacturing 27
oil 105
  blown 107
  drying 106
  drying capacity 106
  non-drying 106
  semidrying 106
oil length 107
olefin type 135
oligomerisation 159
one-to-one crosslinking 118
orange peel 31
original equipment manufacturing 27
overstoving 43
oxygen inhibition 163, 165, 174, 163

P

painting
  electro 25
  purely electrostatic 27
paint materials 18
paint polyisocyanate 130
paint, solids rich 145
paraffin mirror 174
peroxide, for UP curing 173
photo-crosslinking 164
photo-initiator 51, 161, 163
  concentration 163
photon energy 160
physically drying coating materials 12
polyacrylate 151
  non-polar 155
polyacrylopolyol 131
polyacryloyl resin 128
polyanhydride 5, 127
polyaziridine 129
polyester acrylate 164
polyesterpolyol 131
polyester powder coating, curing
  with blocked polyisocyanate 150
  with glycidyl crosslinkers 150
  with glycoluril resin 151

with hydroxyalkyl amide 150
with polyuretdione 150
polyester
semicrystalline 156
unsaturated (UP) 171
polyether acrylate 164
polyisocyanate 125
aliphatic 130
aromatic 130
blocked 133
polyketimine 127, 138
polymerisation 141
cationic 159
polymerisation
free-radical 85, 157
polymer network 60
homogeneous 62
polyol
hyperbranched 119
polyurethane, acrylated 170
popping 49
pores 92
post cure 170
primer 174
active 174

Q

quantum yield 161
quartz UV 51

R

radiant power 46
radiation
dose 53
curing 50, 163
curing, free-radical 160
curing of UP 173
electromagnetic 45
pyrometer 44
radical combination 109
rate constant 83
reactant 81
reaction
bubbles 82
chemical 70
competing 81
Diels-Alder 86
mechanism 81
order of 82
pericyclic 86
polar 84
reactive diluent 12, 164, 169, 171
reactivity control 122
reflector parabolic 53

S

sagging 29, 94, 145, 154
limit 40
salt crosslinking 60
semicrystalline polyester 156
sensitizer 168
shear
thickening 15
thinning 15
siccative, for waterborne paint 112
silicone resin 128
SiH-functional 128
vinyl 128
singlet state 160
skinning 91
softening 64, 144
soft resins 95, 69
sol 144
fraction 62
sol-gel process 145
solids content 12
solubility parameter 70
diagram 65, 72
three dimensional 71
solubility region 72
solution, aqueous 74
solvent
bound 69
effect 132
free 69
influence on T_g 65
-resistance 22, 62
retention 90
selection 92
volitility 68
spreading constant 34
Stefan-Boltzmann law 46
Stefan's constant 46, 47
stoving duration 123
stoving powders 30
styrene 171
styrene-butadiene dispersions 111
sulphonium salt 167
surface
curing 163
energy 33
synergist 161

T

tangle interpenetration 92
termination by combination 158
tetraethyl silicate 144
T_g
  dependence on molecular weight 65
  influence of solvents on 65
TGIC = triglycidyl isocyanurate 149
thioxanthone 161
thixotropic agent 40
thixotropy 15
through-drier 111
titanium dioxide 161
  transmission of 161
transmission, degree of 48
tribo method 31
triplet quencher 163
triplet state 160

U

ultraviolet radiation 50
under-crosslinking 121
understoving 43, 123
urea
  binding 131
  group 121
urethane
  acrylate 164
  formation, mechanism of 130
  group 115
UV
  binders, waterborne 170
  curing, ionic 167
  curing under CO_2 166

V

vacuum coating 29
vacuum UV 51
valence angle strain 136
Van der Waals forces 98, 99
vapour
  curing 132
  pressure 67
  of a polymer solution 72
vinyl silicone resin 128
viscosity characteristic 154
volume shrinkage 155, 165

W

water 64
  -binding capacity 100
  evaporation 87, 88
  mountain 76, 94, 98
  retention 98
  siccative for 112
  -solvent mixture 94
  waterborne paint 50, 68
WLF equation 122, 153
wrinkle finish 112
wrinkling 112, 165

X

xanthone 161
X-rays 167

Y

yellowing 174
yield point 15

Z

zinc complex 129
zinc-dust paint 144
zirconium complex 129